Contents

Acknowledgements

This book was commissioned by Duncan Clark and Mark Elling-ham of *Green*Profile. I am enormously grateful to them. Duncan also carefully edited every page and his influence on the final text is extensive. It was a privilege to work with him. Any mistakes are, of course, mine.

My wife, Professor Charlotte Brewer, and the rest of my family put up with months of sustained surliness. I cannot thank them enough for their gentle tolerance of my bad behaviour. Charlotte also read and commented on the entire work. Her interest in some of these technologies helped keep me enthused.

Christopher Whalen did the picture research with his usual efficiency, orderliness and skill. My father, Peter Goodall, assisted in research throughout the book's gestation. He found much of the background material for the book's ten chapters and pointed me in the right direction on countless different occasions. Every morning a new set of interesting articles and commentaries would arrive by email helping to keep me fully briefed on the increasingly rapid developments in the science and practical technologies covered in this book.

I am grateful also to the science teachers of my secondary school, St Dunstan's College, Catford, London, for their rigour and for their devotion to the task of instilling a lifelong interest in practical science in their pupils.

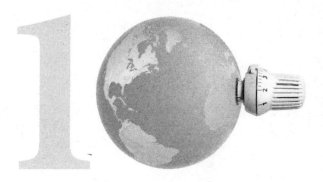

Ten Technologies to Fix Energy and Climate

'Goodall's book contains much of the information you'll need to follow the energy debates in years to come', *The Herald* (Glasgow)

'Rewarding and essential, *Ten Technologies* combines rigorous research and an accessible tone.' BBC Green

'Books about climate change can be depressing, so three cheers for Chris Goodall's latest, *Ten Technologies,* for its positive, pragmatic message. Best of all, you don't need science qualifications to understand it. Written in an accessible, engaging style … Small wonder, perhaps, that this timely, thoughtful book was recently chosen as one of the *Financial Times* Books of the Year. It provides some fascinating, carefully analysed insights into where we might go next.' *Oxford Times*

CHRIS GOODALL is a writer and broadcaster on climate change issues, and editor of the website, Carbon Commentary. His previous book, *How to Live a Low-Carbon Life*, won the 2007 Clarion Award for non-fiction and was described by *New Scientist* as 'the definitive guide to reducing your carbon footprint'. He spends his free time turning off household appliances at the mains socket.

Author's note

Units of energy

This book uses the kilowatt as the base for its descriptions of the power of the various technologies. A kilowatt is the amount of electrical energy necessary to light ten old-fashioned 100-watt incandescent light bulbs, or about a third of the power used in an electric kettle.

One kilowatt of power continuing for sixty minutes is called a kilowatt-hour.

The other units used in this book are megawatts (1,000 kilowatts), gigawatts (1,000 megawatts) and terawatts (1,000 gigawatts). To illustrate the scale of these figures, here are some comparisons. The typical European home uses about 4 megawatt-hours (or 4,000 kilowatt-hours) of electricity per year. A US household uses approximately twice as much. A big fossil fuel power station generates a gigawatt or more. The UK consumes about 350 terawatt-hours of electricity every year and the US about ten times that amount.

The book offers approximate figures for the cost of producing low-carbon energy. To put these figures in context, broad estimates of the price of fossil fuel electricity are as follows. Over the last few years the wholesale price of electric power in the UK has tended to be about 5 pence per kilowatt-hour or £50 per megawatt-hour, although it is somewhat higher at the moment (summer 2008) because of unprecedented fuel prices. The retail price of electricity sold to domestic homes is about twice this figure, covering the cost of distributing the power, maintaining the grid and billing the customer. Prices in Europe are about the same as in the UK, but those in the US are typically somewhat lower.

At times of peak demand or when the electricity system encounters a sudden problem, such as a malfunction in a big power station, the wholesale price of power can suddenly spike upwards to a level several times higher than average, as power stations that are standing idle are offered high prices to persuade their owners to start producing electricity. At present, the wholesale price at which it becomes profitable for the owner of a power station to start producing electricity mainly depends on the cost of coal and natural gas, today's principal fuels for electricity generation.

I am focusing here on electricity because a future low-carbon world will probably use more electric power than at the moment and less oil, gas and coal. Using renewable sources, such as wind or photovoltaic panels, we can generate electricity without producing large amounts of greenhouse gases and policy-makers around the world are keen to encourage a switch away from fossil fuels and towards clean electric power.

A note about greenhouse gases

This book regularly refers to carbon dioxide, the most important man-made climate-changing gas. Carbon dioxide emissions are the major part of the world's greenhouse emissions, which also include methane, nitrous oxide and several other gases which are primarily used for refrigeration or some industrial processes.

A molecule of carbon dioxide consists of one atom of carbon and two of oxygen – hence its formula CO_2. Confusingly, we sometimes talk about the weight of greenhouse gases in terms of carbon and sometimes in terms of carbon dioxide. The crucial thing is that a molecule of carbon dioxide weighs 3.667 times more than an atom of carbon. I have tried to be clear in the text as to whether I am referring to the weight of carbon dioxide or just to carbon.

Introduction

Two or three email newsletters drop into my inbox every week promising spectacular returns if I invest now in green technologies. The overexcited claims of dubious stockbrokers suggest that the battle against climate change will be won as easily as the DVD took over from video cassette. The technologies promoted in these newsletters often have a disturbing reliance on breaches of the hitherto unassailable laws of physics.

This book is more restrained. It does not claim that the world will painlessly escape from the shackles of fossil fuel dependence, quickly and cheaply building a low-carbon economy. But I hope it demonstrates that, however difficult the transition might be, the world has the tools it needs to tackle climate change. The book identifies and explores ten separate ways in which we could significantly reduce emissions or extract large volumes of carbon dioxide from the atmosphere. It also suggests that, once we have successfully switched away from coal, gas and oil, we will find that energy costs are no higher than they are today, and perhaps considerably lower. There are huge technological improvements to come that will reduce the price of low-carbon energy to a fraction of what it is today. The earlier we start a systematic programme of investment in new technologies that don't use fossil fuel, the sooner we will see the costs decline to the level of today's fossil-fuel prices.

The following chapters steer a line between the technophiles who believe that free markets will naturally bring about the growth of alternatives to oil, coal and gas and the growing number of environmental pessimists who think that the world is hurtling towards catastrophe at increasing speed. Most of the technologies discussed are still in their infancy and, although their prospects seem bright, none will advance rapidly without large amounts of risk capital, consistent and expensive support from governments (and therefore also from their electorates in democratic societies), and continued scientific advances. And I hope it goes without saying that these technologies are not a substitute for improvements to energy efficiency across industry and domestic life. The world needs a mix of technical advances and complementary reductions in energy use – including substantial lifestyle changes – if we are to stop and eventually reverse the rise in greenhouse gas concentrations in the atmosphere. Investment now in alternative technologies will also release us from reliance on oil and gas supplies imported from a small number of countries, not all of whom bear the West much goodwill.

Some of the ten technologies in this book will fail, and it is a reasonable bet that a clear majority of the innovative companies that I briefly profile will not even exist in ten years' time. This shouldn't particularly concern us. All that matters is that those technologies which do eventually succeed are rolled out on a massive scale. Even the global warming pessimists should recognise that the world's entrepreneurs, venture capitalists and scientists are devoting unprecedented amounts of ingenuity and hard work to the greatest challenge of our age. This is a global effort, and the following pages look at people and companies in places as diverse as China, the US, Ireland, Spain, Korea, India and Australia. If the world fails to solve its climate change and energy security problems, it won't be because these individuals didn't try hard enough.

The second glass problem

When speaking in public, almost all specialists engaged in the climate change debate offer a positive and hopeful view of the world's ability to tackle climate change. They know that if they say that the situation is too awful and frightening they will lose the sympathy of the audience. Speakers have to be relentlessly upbeat, stressing the capacity of the world to reduce its use of fossil fuels while still improving prosperity around the globe. With a few exceptions, the public stance of climate-change experts is that global warming is within our control, at least for the next few years.

There is often a reception after the speech and the scientist or politician speaker will stay to chat to the people who came to the talk. Glasses of indifferent wine are passed around and the conversation moves to the actions the world needs to undertake to avert the potential of unmitigated catastrophe. I have been to many of these events, and I have noticed the same thing happen on almost every occasion. Winding down after the talk, the speaker sips the first glass and continues to say that the climate problem is within the capacity of the world to solve. But as he or she reaches for a second glass, and the alcohol starts to loosen inhibitions, the speaker begins to offer a less cheerful view. The slow pace of change in attitudes among the world's political elite is witheringly dissected. (I would use the word 'glacial' to describe the rate of progress, but since some Greenland glaciers now move several kilometres a year this adjective is far too generous.) The speaker notes the mounting evidence that the relatively small increases in average temperature we have already seen are having surprisingly dramatic effects. The Arctic will probably have ice-free summers within a decade, major Asian rivers are likely to dry up for several months a year, biodiversity is declining at an accelerating rate, and increases in crop yields are slowing as drought, rising salinity and increasing temperatures affect vulnerable plants. The speaker now says what he or she really believes: the world is not yet ready to make the adjustments necessary to control climate change.

Many in the climate change debate have, understandably, moved on to a metaphorical second glass of wine. They have become deeply pessimistic about human society's capacity to change course quickly enough. They despairingly note that global carbon dioxide emissions appear to be rising at a faster rate than in any of the scenarios envisaged by the reports of the Intergovernmental Panel on Climate Change. The recent economic recession depressed oil and gas demand but the moment growth picks up, energy demand will race ahead. Few countries have begun the process of decoupling the growth in their economies from increasing fossil fuel use. Among policymakers, the pessimists point out, self-delusion abounds. The British government loudly claims success in beginning to stabilise greenhouse gases, for example, but it ignores the emissions from aviation and those 'embedded' in the ever-growing number of products imported from China.

It isn't that the world doesn't recognise that global warming is a problem. When asked, a large majority of people around the world say that human activity is causing changes in the climate. And people are concerned about these changes. In an international survey conducted by HSBC in 2007, 60 per cent of Indians said that climate change was one of their biggest worries. Global-warming sceptics still exist in large numbers but the majority of people, perhaps observing the increasingly obvious evidence from the natural world, accept that rapid and unpredictable climate variations are happening around them. Forests are more vulnerable to fire, storms are increasing in intensity, icepacks and tundra are melting and drought is causing starvation in water-stressed countries.

When individuals are asked whether climate change can be successfully controlled by humans, we see large differences between countries. People in the developing world are much more inclined to believe that the global community can successfully arrest global warming. In India, 45 per cent of respondents said we can control temperatures, but in France the figure was less than one in ten. The inhabitants of rich countries, usually responsible for a disproportionate share of greenhouse emissions, are generally not optimistic about mankind's ability to solve the carbon dioxide problem.

So who's right? Are the attitudes that come to the surface when sipping the second glass of wine reasonable, or are there good grounds for the optimism that is widespread in India, Brazil and China? Is it too late, or perhaps just too technically difficult, to reduce our economic reliance on fossil fuels?

This book argues that there is reason for very considerable optimism. Each of the ten chapters looks at a technology or technique that could reduce CO_2 emissions by at least 10 per cent of the annual world total. All of them are comfortably within our scientific and technological reach. So, to use that ugly phrase, we should be able to 'decarbonise the economy' at an affordable price.

In fact, many of the technologies in this book, such as zero-till farming or improved home insulation, can be implemented today with no permanent increase in costs. They will improve incomes, make agricultural yields more reliable or reduce household expenditure. Other technologies, including second-generation biofuels, and tidal energy, will probably be more expensive than their fossil fuel equivalents for some years or decades to come. But every chapter concludes that with reasonably predictable technological progress we can expect that our energy sources will eventually be no more expensive than at present. Importantly, I also suggest that carbon sequestration – ensuring that CO_2 is permanently stored – is a readily available option, albeit at some cost. This is very good news.

Nevertheless, I don't want to suggest for one second that phasing out fossil fuels is going to be easy. After all, when I give talks on climate change I always refuse the second drink after the speech, for fear I will let my own worries show. We should accept that some of the technologies in this book require the world to make wrenching changes to the way we do things. At the moment, for example, we power our cars with petrol or diesel. The fuel we need is available at many thousands of filling stations at the side of the roads across the world. Liquid fuels are convenient and give us unparalleled flexibility. But from a climate change point of view, it makes eminent sense to move to a world in which we all use electric vehicles powered by batteries charged with energy from renewable sources.

Eventually, battery-powered electric cars will be cheaper and easier to maintain than the dinosaurs of the internal combustion era. But shifting the world's car fleet to running on electrons rather than petrol is not a trivial task. Batteries need substantial improvement in cost, the speed with which they charge, and their capacity to store enough power to drive the car for hundreds of miles. Although the first all-electric sports cars are appearing on the road to excited reviews, this does not mean that batteries will rapidly become the main means of automobile propulsion. We need entrepreneurs and corporations to take huge risks in moving away from petrol. Governments must offer support and fiscal encouragement. Car users will need to be tolerant of the flaws of the first generation of electric vehicles. But once we have got over the humps in the road, everybody will wonder why we took so long to switch to pollution-free, easy-to-maintain, super-efficient battery cars.

Almost all of the other technologies described in this book will go through similar phases: an expensive and inconvenient

An early US wind turbine

introduction; a troubling period in which enthusiasm wanes and improvements appear to be slow; gradual acceptance by sceptical purchasers; and, eventually, a dawning sense that we really can do without the fossil fuel alternative.

There is one particularly easy way to attack my restrained

optimism: point to the experience of the early 1970s. Then, as now, the price of oil had accelerated upwards at a dizzying rate. Governments and companies around the world were eager to rid themselves of dependence on the oil cartel. Research and development (R&D) programmes tried to find the best way of commercialising low-carbon technologies. Many of these R&D efforts went into precisely the same set of technical opportunities promoted in this book. The US government put money into biofuels, the Chinese invested heavily in anaerobic digestion, the UK began research into wave power, and governments in Europe backed combined heat and power plants. They are doing exactly the same today. Unfortunately, however, they are often spending a smaller percentage of our national incomes than they did thirty-five years ago.

In many cases, as the cynics never cease to remind us, the earlier attempts to speed the development of new energy-generating techniques were complete failures. Costs remained high, the technology immature and consumer interest limited, despite the investment of billions of dollars of public money. Generally declining fossil fuel prices in the three decades after the oil shock of the early seventies caused governments to lose interest and most research efforts faded away. It is one of the great ironies of the last few years that some of the scientists involved in the 1970s alternative-energy drive have been brought out of retirement to restart the same R&D programmes that were abruptly shut down decades ago. The people who came closest to finding an industrial-scale technology for making diesel fuel from growing and then crushing algae are back in the labs they left thirty years ago.

Some of the other early pioneers of low-carbon technologies, perhaps conscious of the transience of the interest of government and investors, have decided to move on. Salter's Duck, one of the first wave-power collecting devices, was designed in 1974, just after the oil shortages of the preceding year. It was a genuine advance and its efficiency in capturing the energy in waves has scarcely been bettered since. Professor Stephen Salter, its South African-born inventor, now has the pleasure of watching wave power finally being commercialised. But in the interim, after decades of minimal official interest in renewable power, Professor Salter has switched focus. He and his colleagues are now investigating a technology to increase low-level cloud cover over the oceans. Since low clouds block the sun's rays, Salter's scheme might help limit global warming. This is one of the ten or so 'geoengineering' projects that the more pessimistic among the scientific community are investigating in the hope of dealing with the consequences of the carbon dioxide build-up, rather than trying to avert it. The concluding section of this book looks sceptically at how we might employ some of these schemes in a climate emergency.

Given the rapid fall in enthusiasm for alternative energy two or three decades ago, why should we believe that the current level of interest will be any more persistent? Is it not simply naive for us to think that the nascent technologies of solar power, fuel cells and advanced biofuels will ever be competitive with fossil fuels? Three forces bolster my optimism.

First, many of the technologies that looked good in 1973 but failed commercially have experienced significant and sustained reductions in price since then. Many important low-carbon technologies – most obviously wind power – have come down in cost with almost predictable regularity. Consistent with the well-understood theory of the 'learning curve', manufacturing costs have gone down by a similar percentage every time the number of units produced has doubled. If the experience of almost every single manufacturing industry in the world is any guide, the next doubling of the accumulated volume of wind turbines will reduce costs by approximately the same percentage. Even though the recent headlong rush into wind power caused a sharp spike in equipment prices, underlying costs will continue to fall as manufacturers gain knowledge of how to build turbines more cheaply. By contrast, as the world pumps more oil from its increasingly depleted stocks, the price will almost certainly tend to go up. Although we continue to see far more money invested in oil exploration than in the development of alternative energies, the number of barrels found per million dollars spent is still declining. At some point, perhaps soon, the financial returns to investment in low-carbon technologies will exceed those from drilling for oil and gas. At that point, costs of new technologies are likely to sharply dip.

Second, the world is now concerned about climate change. With the exception of a few peculiarly far-sighted scientists, no one was worried about global warming in the early 1970s. This means that low-carbon technologies are more likely to garner the long-run support and subsidy that they need.

Third, opinion leaders around the globe have an increasingly strong sense that the world is beginning to run out of minerals, or at least failing to keep up with the increase in demand. The recent simultaneous increases in the prices of metal ores, fossil fuels, and fertiliser sources such as phosphate created awareness – not before time – that the globe cannot be indefinitely mined. It now seems painfully obvious that future economic growth cannot be based on unlimited supplies of raw materials, available for little more than the cost of extracting them from the ground. This is a sharp about-turn from the attitudes of even five or ten years ago when pessimism about the long-run availability of raw materials was confined to a few inveterate doom-mongers. Even if we did not need to reduce fossil energy consumption for climate change reasons, there are compelling reasons to find ways of living without continuous recourse to scarce and increasingly expensive materials extracted from a thin layer of the earth's crust. Low-carbon energy sources have the advantage of working with the grain of this important change in the *zeitgeist*.

Obstacles to the ten technologies

Just because a technology is good and its financial advantages clear does not mean that it will be seamlessly and quickly incorporated into widespread use. In each of the ten chapters of this book, I try to note the chasms that have to be crossed – technical and financial – before we see truly widespread adoption of the most promising of low-carbon opportunities. I have done this partly to rebut the accusation that the chapters of this book are little more than public relations on behalf of the new industry on which I am commenting. I also suspect that we will go through several cycles of elation and disappointment before the full outlines of a low-carbon society become clear. It is better to recognise early that the road is not going to be easy.

Loss of convenience

One of the main obstacles to the adoption of new technologies is the ubiquity of the existing infrastructure that enables us to use fossil fuels cheaply and conveniently. Thousands of billions of dollars have been spent building natural gas pipelines and storage tanks, electricity distribution grids, huge coal- and gas-fired power stations that operate safely and reliably, mostly with few hours each year of unscheduled maintenance, and networks of oil refineries and filling stations. It won't be easy to switch away from the pipes, wires, buildings and machines that have been so expensively built up over the last century or so and have served the inhabitants of prosperous countries so well.

It is true that some alternative energy products can be fitted into the existing infrastructure. Cellulosic ethanol can, for example, be mixed with conventional gasoline without requiring new cars, filling stations or oil refineries. But other technologies require new distribution systems. Wood-based community heat and power plants, for example, rely on the installation of hot water pipes around urban areas. The large companies at the centre of the fossil fuel economy – electricity generators, oil companies and pipeline operators – have the human and financial resources to invest in projects of this scale. Few institutions have the financial capacity or skills to do the same in the low-carbon world.

Moreover, our lives are currently structured around instant and consistent access to energy. For example, the electricity system in advanced countries offers nearly universal access and reliability. Even a brief power loss in a developed nation can prompt startled front-page newspaper headlines. It would be naive to expect that low-carbon technologies could match this reliability, and replace all the other advantages of fossil fuels, within a few short years. We will go through periods when the new technologies fail, provide only intermittent supply, and cost more than their fossil fuel equivalents.

Archive films from around 1900 show many spectacular, and often very funny, failures of prototype airplanes to get off the ground. Low-carbon research will throw up similar disasters. The last few years have offered several good examples. In one case, algae grew so fast that they overwhelmed the inlet pipe at a power station where they were being tested as a way of capturing carbon. A wave power collector fell to the bottom of the ocean. A turbine blade fell off a tidal power device. Such failures provide an opportunity for sceptics to brief the press with the message that renewable technologies will never replace the electricity, gas and petrol so ubiquitously provided at present. If we're to tackle climate change, however, we must take a leaf out of the Wright brothers' book and not be deterred by early difficulties.

Resource shortages

Another issue facing some of the technologies profiled in this book is a shortage of key components. In 2008, wind turbines suddenly became in short supply because the rapid increase in demand left manufacturers and their suppliers unable to make enough of these surprisingly complex pieces of equipment. After decades of decline, turbine prices rose sharply. Marine energy development is being held back by a worldwide shortage of the specialist ships that can carry out operations on the sea bed. At the other end of the spectrum, the growth in the number of electric cars on the road may be limited in the longer term by shortages of minerals, such as lithium, needed to make their large batteries. Similarly, some types of fuel cells, though not the ones promoted in this book, are reliant on supplies of currently eye-wateringly expensive platinum, and many solar photovoltaic companies compete for the limited world supply of ultra-pure silicon (although the shortages of 2008 turned into the enormous glut of late 2009).

As we build up the low-carbon economy, we will experience repeated bottlenecks, periods of dramatic oversupply and painful interruptions that temporarily check progress. This is not a persuasive reason for holding back the development of alternatives to fossil fuels. It simply means we will experience a slower and more painful transition than we might have hoped for.

The need for scalability

The environmental movement sometimes conflates its concerns about climate change with its deeply held dislike for many aspects of the modern economy. Eco-activists often rail against impersonal and amoral multinational corporations, the gigantic and unresponsive public utilities that dominate electricity and gas supply, and the political sway of fossil fuel interests. Partly as a result, well-meaning environmentalists and green politicians often prefer to support technologies operating on a small scale. They back subsidies for solar panels on houses, small wind turbines and wood-burning home heaters. There is nothing inherently wrong with this, except that these technologies are far more expensive for each unit of carbon saved than their full-scale equivalents. For this reason, micro-renewables will not stop climate change.

Take Germany's subsidy of small-scale solar photovoltaic installations. This is an unprecedentedly generous initiative. It costs about €5bn in annual payments but solar still generates less than 1 per cent of the country's electricity. More importantly, perhaps, the subsidy scheme has sucked in much of the world's supply of photovoltaic panels and put them on roofs at latitudes where they were likely to generate less than half the electricity they would have done in, say, southern Spain or Mexico. Although Germany's subsidy has helped build the businesses of the main Californian and Chinese photovoltaic panel manufacturers, it also substantially pushed up the price of silicon to the rest of the world for several years.

We need to apply the ten technologies in this book in ways that maximise their benefit, and this generally means large-scale implementation. Electricity from a wind turbine attached to a house in the centre of Europe might cost five times as much as from a properly sited large wind farm in a good location on the Atlantic seaboard. Does it really make sense to subsidise the smaller turbine? Probably not: it may not make us particularly comfortable, but the low-carbon world may have to be dominated by companies as large as today's oil and electricity companies. We need corporations that can invest tens of billions of dollars every year in huge projects in every country in the world. To fight climate change, we must use the strengths of global capitalism, not pursue an unwinnable battle to use the threat of global warming and energy insecurity to alter the way the world economy works.

As the German example shows, it's also important to put

alternative energy sources in the most appropriate locations. It makes sense to focus on battery-driven cars first in countries where typical driving distances are short and petrol is expensive. City states such as Singapore and small countries such as Israel are good examples. Success will be much easier to achieve than in places where people need to drive long distances and fuel is relatively cheap. Soil carbon improvement programmes should be concentrated in those countries with large expanses of carbon-poor soils and well-established educational infrastructures to help farmers understand the merits of different grazing practices. Carbon capture and storage will work best in those countries with abundant deep saline aquifers. These are all obvious points, but in their zeal to be thought to be doing useful things, governments, companies and individuals have sometimes been distracted by the comforting acceptability of micro-initiatives rather than focusing on technologies that can be implemented at truly gigantic scales across the world.

Individual countries need to assess which technologies are most relevant to their particular circumstances and focus their limited resources on these opportunities. Cloudy Britain is wasting its money subsidising the installation of solar hot-water units on domestic homes when it could sponsor R&D into exploiting the country's awesome resources of marine energy. By contrast, China should continue to concentrate on small-scale biogas digesters, highly forested countries like Sweden and Canada on wood-based community heat and power plants, Spain can commit to solar energy, Denmark to maintaining its unrivalled expertise in wind power, and Australia to soil improvements. Given the importance of the car in US society, it also seems to make sense for the country to continue to lead the world in the development of second-generation biofuels, such as cellulosic ethanol.

Focus by individual countries on two or three of the ten technologies in this book is much more likely to result in commercial success and continuing improvements in cost and usability. Scattered, unreliable and inconsistent support may actually be counterproductive because it will divert resources from more appropriate objectives. The battle against global warming should not be a game of roulette with countries tossing a few chips towards random technologies. Research and development, public investment, and tax incentives need to be thoughtfully targeted.

Linkages between the ten technologies

It may seem inconsistent to suggest that countries should focus on the technologies which are likely to be most appropriate for their circumstances, whether economic or geographic, but then stress the need to understand the close relationships between each of the ten proposed solutions. The point I am trying to make is that it will make sense for governments and companies to invest in a smaller number of promising opportunities but that countries will need to deploy the full spectrum of technologies in order to ensure an energy supply that is almost as dependable as fossil fuels are today in rich countries.

Symbiosis

Fossil fuel energy has gained its dominance partly because it is so utterly reliable. With the possible exception of wood-burning power stations and liquid fuels made from biomass, no renewable electricity technologies offer quite the same degree of consistent availability. If we are to completely run the electricity-generating systems of large countries without fossil fuel power, which is a much less far-fetched idea than most people assume, we will need to find ways of ensuring that each new technology buttresses, rather than undermines, the others.

Take wind power. The unpredictability of the winds means that individual grids, whether regional or national, cannot easily accommodate more than 10 or at most 20 per cent of wind-generated power. So the growth of wind turbines needs to be complemented either by power storage, such as in car batteries, or by electricity sources that can be turned up and down quickly, such as fuel cells or wood-burning power stations. If the wind stops blowing, batteries in electric cars that are plugged into the electricity system, perhaps in domestic garages, can be gradually discharged to provide back-up power, or fuel cells that generate electricity for the home or business can be remotely commanded to increase their output rate.

Similarly, solar energy, which will only ever deliver direct electricity 12 hours or so a day, will need large-scale storage systems. And tidal energy, which peaks according to a predictable cycle but at varying times of day at different points on a coastline, should be widely geographically dispersed to make sure that its contribution is as consistent as possible.

One conclusion that I've taken from my work researching this book is that energy supply in each country will probably need to be carefully planned by a central authority. The free market will be very useful in deciding which potential technical innovations offer the best opportunities, but it will probably not give us the tight integration of various complementary technologies that the world needs. This point is forcibly made when we look at the likely impact of encouraging the growth of nuclear power on the incentives to invest in low-carbon energy sources.

To an extent that seems not to be appreciated by policymakers, nuclear energy is the enemy of renewable sources of power. Nuclear plants cannot be switched on or off at short notice: to be cost effective they must run 24 hours a day. Renewable sources generally also benefit from running all the time that their power source is available. Wind, solar and marine power sources are all expensive to build and cheap to operate, so it makes clear financial sense to run them for as many hours as possible.

So nuclear power stations and renewable energy sources such as wind turbines are in direct competition. If a large number of nuclear plants are generating power every day of the year, there may not be much need for other forms of supply. The UK government is talking of encouraging utilities to build new nuclear power stations that would be able to supply all the electricity the country needed at periods of minimum demand, which is usually at about 5.30 a.m. This means that when demand is low, wind turbines (and all other renewable power sources) would need to be disconnected from the electricity grid, and the owners of these assets would not be paid for their electricity. This makes investment in wind more risky than it would otherwise be. Unless large amounts of power can be exported to other countries freely and at reasonable prices, a large nuclear industry is incompatible with encouraging major investment in wind or any other sources of renewable electricity.

It may remind of us of the old days of central planning in Eastern Europe, but renewable generators need to be given a clear and quite precise promise about how much other generation capacity is to be constructed.

The importance of land use

We can eventually obtain most of our electricity from renewable sources directly or indirectly powered by the sun. Wind energy, wave power and solar technologies all convert power that originated in the sun's nuclear reactions. The chapters on these technologies give figures for the percentage of the world's electricity demand that each can comfortably provide. The percentage of the earth's surface that will be needed is not large.

But we will also need to use some of the sun's energy that has been captured in plants and trees. Biomass, the technical expression for energy sources created through photosynthesis, is going to be an increasingly important source of our electric power and liquid fuels. The material on cellulosic ethanol (see Chapter 7) shows how wood and straw will be converted cost-effectively to a petrol substitute. But cars are prodigious users of energy. A typical small European vehicle consumes about 1,500 litres of fuel a year. Even with efficient new technologies that convert wood and straw into ethanol, one such car will need the yearly cellulose output from a quarter hectare of land. If this hectare was good grain land, it could have produced enough wheat to provide the food needs of twelve people for a year.

One of the great issues the world faces is how it decides to allocate land between the various competing uses. Four chapters of this book assess technologies that in one way or another use the resources provided to us by the processes of photosynthesis, the turning of light into plant growth. We will need to devote land to growing woody biomass for ethanol (not to be confused with the foolishness of using foods for biofuels) as well as for the fuel in combined heat and power plants. Chapter 10 shows that we can also productively sequester carbon by digging charcoal made from wood and plant matter into arable soils, and the final pages of the book look at taking greenhouse gases from the air by improved techniques for pastoral agriculture and from reforestation. (The production of algae in carbon capture plants might also remove land from its alternative uses in agriculture. But because algae cultivation uses so little space, this is not a major concern.)

A titanic struggle is in prospect. The rich world will want to use land around the globe to deliver the biomass resources for its car fuel and for electricity generation. And it has the financial strength to achieve its wishes. Even at the historically high grain prices of the last few years, arable land is potentially worth far more growing biomass for a cellulose-to-ethanol plant than for growing wheat. At the same time, the world's people need the land to grow food. Simply put, 600 million cars are competing with the food needs of 6 billion people for the products of the land. If oil goes up in price again, it will encourage more farmers to turn to growing biomass for conversion to ethanol, which will indirectly increase the price of other agricultural commodities.

The conflict between food and fuel for the world's productive land might seem to make it impossibly difficult to devote increasingly large areas to energy crops and to woody matter that can be used as a source of charcoal for 'biochar' (see Chapter 9). Even without factoring in the impact of using land for climate change mitigation, US Department of Agriculture projections envisage growth rates in agricultural production dipping below the rate of increase in world population. This will erode the substantial gains in food availability that the world has seen in the last few decades. At first sight, it looks as though we cannot reconcile the need for more food production with the requirement that we devote perhaps 10 or 20 per cent of usable land to increased biomass production. But perhaps this pessimistic conclusion is unwarranted. In the last two chapters I try to suggest that just as we have mined the globe for fossil fuels and minerals, we have also mined the soil, degrading its ability to grow the food the world needs. This degradation has significantly reduced the amount of carbon held in the soil. We need to reverse this process, working to gradually improve the agricultural productivity of marginal lands. Sequestering carbon in the soil through the use of charcoal (Chapter 9), or by improving management of grazing land and forests (Chapter 10), means that much of the land around the world can be made much more productive, easing the conflict between the need to produce more food and to create more usable biomass. Improving soil health has two beneficial effects: improved agricultural yields and net carbon extraction from the air.

Some scientists have proposed elaborate machines to take existing atmospheric carbon dioxide out of the air. Global Research Technologies' elegant solution is discussed in the chapter on carbon capture and storage. Like these scientists, I believe that the world does need to find low-cost solutions that directly reduce carbon dioxide levels by taking it out of the air as well as investing in technologies that give us abundant energy without adding greenhouse gases to the atmosphere. The hypothesis in this book is that soil improvement is the cheapest way of achieving net extraction of existing carbon dioxide stocks in the air. It also has the manifest fairness of delivering most of the benefit to people in low-income countries, perhaps making this proposal more politically palatable to the global community.

How difficult can it be?

Sometimes I hear people say that climate change is an impossibly difficult problem and that advanced societies shouldn't even bother to try to avert future warming. Instead, they claim, we should try to adapt to temperature and moisture changes as they occur. Perhaps surprisingly, those who make this claim are often among the people usually found praising the ability of modern capitalist economies to adapt flexibly and quickly to any challenges that arise. It has never been clear why this group are so frightened of the temporary disruption in energy costs that a low-carbon world would endure, while at the same time proclaiming the wider resilience of the free-market economy. Why is it that enthusiasts for the free market are so sure that business is capable of dealing with most challenges but unable to adjust to the impact of switching from fossil fuels to renewable energy sources? This is a vitally important point: widespread adoption of carbon-reducing technologies is going to be very disruptive but the great strength of the modern capitalist economy is an almost astonishing resilience and flexibility. Free-market economies have

many flaws, but they are impressively successful at finding ways around technological problems.

Let's look for a second at the scale of the task we might be setting ourselves if we act now. Advanced economies typically spend about 5 per cent of their gross national product on energy – one in twenty dollars of their national income. The figure is slightly higher in the US at about 7 per cent. An extremely pessimistic view might be that a portfolio of carbon reduction measures taken from the ten technologies discussed in this book could temporarily double this percentage. This increase could conceivably persist for a few years before technical advances improved efficiencies and reduced the cost. So for five or ten years, the need to avert potentially catastrophic climate change might require the rich world to spend 10 per cent, rather than 5 per cent, on fuel and energy costs.

Would this change our societies beyond recognition? Would it impose an impossible burden on this generation or the next? Of course not. Dealing with the threat of climate change could conceivably cause a maximum cut of 5–7 per cent in our living standards for one decade. To put that in context, the last few years have seen the greatest increase in material prosperity ever known; 5 per cent is less than the growth in global GNP from 2004 to 2007. So it seems safe to say that we can accommodate all the costs of dealing with emissions reductions with relatively little disruption to our way of life. We would, in effect, be abandoning about two years' economic growth. Moving from coal-fired electricity generation to wind and solar power may well be difficult, but it is not going to cause catastrophe to the modern economy, despite what the global warming sceptics say.

Capturing the wind

Clean power that's more reliable than you'd think

Adam Twine tends an organic farm not far from Oxford in southern England. The surrounding areas are flat and low-lying, but Adam's land occupies a small and windswept plateau. Down below in the far distance the cooling towers of Didcot power station are the most visible marks on the landscape. Didcot is a decaying coal-fired generator due to close in a few years because it cannot meet the latest European emissions regulations. Adam's fields are not ideal for wind power – central England has far lower speeds than the western coasts and many other regions around the world – but he decided in the mid-1990s that he wanted to build a wind farm owned by the local community, sited as a perfect contrast to Didcot, the single largest source of carbon dioxide in the prosperous southern heartland of England.

As with many larger-scale wind projects, the struggle to get the turbines constructed was a long one. It took the best part of a decade just to get planning permission. Although Didcot's six huge cooling towers and the multiple pylon lines trailing away from the power station have already had a huge impact on the landscape, local resistance to the visual effect of the turbines was fierce. When the approval was obtained, a protracted process of fundraising began. By the time the capital was finally raised, a worldwide shortage of components had pushed the prices of turbines up 30 per cent, so more cash was needed. With a few grumbles and support from the Cooperative Bank, the shareholders obliged, raising the final

*The turbines on Adam Twine's land
with Didcot power station in the background*

instalment with a few days to spare in the spring of 2007.

In February 2008 the wind farm finally started producing electricity. Five 1.3-megawatt turbines now rotate sedately (and very quietly) whenever the wind blows. Over 2,000 people own shares in the development. Some invested because of a passionate belief in renewable energy; others because the venture promised good financial returns. As an enthusiastic member of this cooperative, I regularly scan the local forecasts eager to check that the expected winds are as high as predicted in the initial business plan. So far, Adam Twine says that the output from the wind farm has more than delivered on the promises made (and I got my first dividend payment, as promised).

As Adam's farm shows, new wind farms are already good investments in many parts of the world. The best returns come from buying the largest possible turbines, installing just one manufacturer's product at the site and then building as many as possible in the local area. This reduces the costs of connecting the wind farm to the grid and minimises the amount paid for yearly maintenance. In countries such as Portugal, the largest wind developments are now obviously competitive with fossil fuel sources of electricity. BT, the UK's largest telecommunications company and the user of more than half a per cent of the country's electricity, says that its wind turbine construction programme, planned to provide a quarter of its needs, is easily justified to its shareholders as making good financial sense.

The years 2006–8 saw a sharp rise in the price of turbines as the steel for the supporting column and copper for the turbine wiring suddenly cost far more than ever before. The rapid growth in demand for turbines also caused production bottlenecks for some of the 8,000 components in a typical turbine. The shortages have now eased and costs have fallen. The long-run trend in turbine is downward, with expected costs falling from about $1,200 per kilo-watt of generating capacity down to perhaps $800 by 2013 – roughly equivalent to the capital cost of a new gas-fired power station. (The full cost of Adam Twine's wind farm came to almost twice today's average, inflated by the relatively small size of the development and the expensive struggle to get permission to build it.)

As their critics never tire of pointing out, wind turbines do not generate their maximum power all the time. They only produce their full output when the wind is blowing strongly. But not *too* strongly: above a certain wind speed, the machines shut down to avoid the blades rotating too fast and damaging the turbine. Averaged across the year, a 2-megawatt turbine in a reasonable location will typically produce only about a third of this figure – about two-thirds of a megawatt. The wind farm on Adam's land will probably generate about 13 gigawatt-hours in its first year. This sounds an impressive figure, but the old dinosaur of a coal-fired power station down the road at Didcot will produce the same amount of electricity in a busy afternoon. It would take nearly a thousand wind farms the size of Adam Twine's to replace just one power station of this size. Even the huge 130-turbine offshore wind farm at Cape Cod in the US – which was finally granted planning permission in late 2009, after a decade-long battle – will replace only about a tenth of one generating capacity of the largest US coal-fired power plants.

Given their relatively small output and inconsistent perform-ance, are wind turbines a genuinely useful tool in the fight against climate change? The answer to that question is an emphatic yes, and this chapter explains why.

The power of the wind

Wind arises from variations in atmospheric pressure between dif-ferent parts of the world. Air tends to flow from high- to low-pres-sure areas, with the speed of the wind depending on the gradient between the cells. The ultimate cause of these pressure differences is the differential amounts of solar heating across the globe. Wind can therefore be thought of as an indirect form of solar energy. A small fraction of 1 per cent of the light and heat energy received by the earth gets turned into the moving, or 'kinetic', energy of the wind. We can capture this energy using windmills or wind turbines that slow down the speed of the air, transferring power to the rotation of the blades.

A wind turbine can be thought of as the opposite of an elec-tric fan. A fan uses an electric motor to turn the blades when the electricity is turned on: a turbine does the reverse. The rotating arms turn gears, which then rapidly rotate an electrical conductor, usually a dense mesh of copper wire, inside a powerful magnetic field, inducing electricity to flow.

A wind turbine cannot capture the full power of the wind. The theoretical limit is just under 60 per cent of the energy in the flow of air. And the amount of electricity generated by the rotation of the blades is only equivalent to about 70 per cent of the energy captured, even in an efficient new turbine. Even with these disadvantages, wind is still a very productive source of electric power, comparing favourably with solar photovoltaic panels, which turn less than a fifth of the energy they receive into electricity.

The secret of wind's success is the sheer mass of moving air that passes through the rotating blades of a turbine. Air may seem almost weightless to humans, but each cubic metre actually weighs more than a kilogram – as much as a bag of sugar. A strong gust consists of air moving at perhaps 40 miles an hour, or 17 metres per second. This means that every second over 20 kilograms of air pass

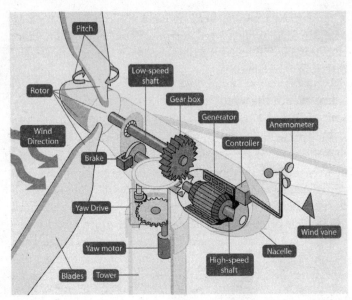

The main components of a wind turbine

through each vertical square metre. This motion contains a substantial amount of energy, with the power in the wind proportional to the cube of the speed of the air. In other words, a wind turbine in a 7 metre per second air flow will generate almost 60 per cent more power than one in a breeze of 6 metres per second. (This is why it is so important to choose windy sites for turbine locations.) At 7 metres per second, which is little more than a gentle breeze, the motion of the wind contains about 200 watts of power per square metre. This is less than the full power of the midday tropical sun, which delivers more than a thousand watts in the same area, but wind is easier to convert to electricity and will often blow for the full 24 hours in the day, not just during daylight hours.

Of course, the amount of power that can be captured by a wind turbine is also linked to the area of the circle swept by its blades. The very biggest new turbines have arms that are over 60 metres long: in a 40 mph wind about 200 tonnes of air will pass through the blades' circle every second, with a usable energy of more than 2 million watts. By comparison, a tiny domestic wind turbine with blades 1 metre long covers just over 3 square metres, capturing up to 600 or 700 watts. Somewhat counterintuitively, the large turbine doesn't sweep sixty times the area of the domestic turbine; it covers over 3,000 times as much.

Wind turbines will probably stop increasing in size soon. There's talk of giant 7-megawatt turbines for offshore installations, but the limit may be reached at 5 megawatts. The problem is that longer turbine arms, while providing more power, are also subjected to more stress. As the arm swings downward, it stretches under its own weight; as it swings upwards, it becomes fractionally compressed. Repeated millions of times a week, this will destroy all but the strongest and most flexible materials – and the longer and heavier the blades, the greater the forces they need to withstand.

Wind's growing importance

Only about 1 per cent of world electricity demand is met today by wind, but the figure varies enormously around the world. Some areas of Germany generate more wind energy than their total power needs. Almost 20 per cent of Danish electricity comes from wind and almost as much in the Canadian province of Prince Edward Island. These high levels of wind power can be accommodated by the local electricity companies because of their freedom to export excess power when the wind is blowing hard, and import electricity when the air is calm. Denmark's access to Norwegian hydroelectric power is particularly important.

Across the world, the US and Spain are adding the largest amount of new generating capacity every year. Wind energy in India and China is also becoming increasingly important: in China, the amount of wind generation has doubled every year over the last three years. By 2015, China may have 50 gigawatts of wind capacity, or about half today's global total. In developing countries without a national electricity grid, wind power combined with large batteries will often represent the cheapest reasonably reliable way of generating power for small communities.

There's no shortage of windy sites left to exploit. One study put the average power in the global winds at any one moment as about 72 terawatts – around thirty times the world's electricity requirements, or 10,000 times more than the wind power we currently generate. And this estimate only includes sites with average wind speeds above 15 miles per hour, a level usually only met at coastlines or on the top of hills. No one pretends that this entire potential can be captured, but we will be able to use wind to provide a good fraction of total world energy needs, and we can expect the rapid growth of the industry to continue for several decades.

Some of wind's growth is being pushed by subsidy schemes. Spain's 30 per cent annual increase in wind power is propelled by price guarantees for the electricity that the turbines generate. But most experts now think that onshore wind turbines are close to competitive with traditional forms of electricity generation, at least in windy locations. It is surprisingly difficult to assess fairly whether wind is cheaper or more expensive than gas turbines or coal-fired stations. It critically depends on assumptions about inflation, interest rates, and how long the turbines will last. And, of course, it depends on the price of fossil fuels. Nevertheless, the long-run trend is unambiguous: wind is going to become a relatively inexpensive provider of power, and if fossil fuels continue to increase in price, this advantage will become more pronounced. Wind generation has its problems and complexities, some of which are discussed later in this chapter, but it provides us with the best possible example that technological progress, heavy investment and government help can push a new technology forward. The cost of wind power has probably fallen by a factor of ten in the last twenty-five years and we can reasonably hope that some of the other infant technologies in this book will improve to a similar degree.

Wind provides a little less than 4 per cent of the European Union's electricity today, four times the average for the world as a whole. The trade body for European wind thinks that this figure will rise to about 13 per cent in 2020, and continue to increase rapidly thereafter. This would mean installing around 10 gigawatts of wind capacity each year over the next decade or so. That equates to thousands of new turbines annually, but since the new capacity installed in 2008 alone was almost 9 gigawatts, the target seems to be well within reach.

Getting to this level will require capital expenditure of almost

€10bn per year, even if turbine prices fall as expected. For this money, Areva, the main European nuclear construction company, says that the continent could have two or three atomic power plants, although the final pages of this book cast some doubt on whether nuclear power can be delivered at this price. Three nuclear stations would have a capacity of almost 5 gigawatts and typically they would operate at full power more than 90 per cent of the time. Even in windy offshore locations, wind turbines with a total maximum power of 10 gigawatts will provide at most 40 per cent of their rated power, or just 4 gigawatts. So the maths is quite simple: capital investment of €10bn a year would buy Europe more low-carbon energy if invested in nuclear power than in wind.

However, once constructed, wind energy is close to free – the cost of annual maintenance is usually a small percentage of the value of the electricity generated. Nuclear fuel is not very expensive but nevertheless its cost helps equalise the price of the two forms of electricity generation. Add in the unknown costs of indefinite safe storage of nuclear waste, and wind seems only a little more expensive than nuclear energy. It will eventually be cheaper, particularly for turbines on windy coastlines.

So why do the big power companies, in the UK and in some other countries, still hanker after more nuclear plants? The reason is probably that nuclear plants are large, centralised, and work around the clock. It is far easier for a big corporation to manage a small fleet of nuclear plants than it is to control a network of thousands, perhaps tens of thousands, of turbines spread across large numbers of sites. The nuclear option also makes it simpler to match electricity supply with customer demand: if the company relies on erratic wind supplies, it will frequently be forced to buy in alternative sources of power at unpredictable prices. In other words, nuclear generation works well for the big companies that dominate power generation in most countries even though it will probably not deliver lower costs than power generated from large land-based wind farms.

Low and predictable running costs also help wind compare well to fossil fuels. Once the turbine is placed on top of its tower, virtually free electricity will be generated for the next twenty-five years or so. By contrast, the world now assumes that coal, gas and oil are going to get increasingly expensive. So investment in wind mitigates the burden from increasing prices of other fuels. Wind has an additional advantage, too. Because its fuel is free, the turbine owners will generally be able to sell their electricity at all times and make a profit. By contrast, the main fuels for power stations – gas and coal – can swiftly vary in price in relation to each other. The hundreds of millions of pounds invested in a coal-fired generator may produce nothing for months if the price of coal rises too high compared with the cost of gas. Didcot coal-fired power station, a few miles from Adam Twine's wind farm, sat idle over much of the winter of 2007–8 because of an unpredicted spike in the price of coal meant that it was uneconomic to run the plant.

Reliance on fossil fuels has a real cost to the economy if consumers and manufacturers cannot guess what energy prices are likely to be months, years or decades in advance. One of wind's primary but often underestimated virtues is that it delivers electricity without such financial volatility. The output of a wind farm may be uncertain, but the cost is not. And, of course, wind power is independent of political intervention – countries that invest in wind are less reliant on the two or three states that provide much of the world's natural gas.

Kick-starting an industry

Denmark began to build substantial numbers of wind turbines in the 1990s, and became the first nation to generate a significant fraction of its electricity from this source. The early years were characterised by the installation of hundreds of what are now considered very small turbines. The owners were local cooperatives and farmers, and these pioneers allowed Denmark to develop a world-leading wind-turbine industry.

In recent years, developers have replaced these small entrepreneurial groups, constructing much bigger farms using bigger turbines. But the community approach is still important. Take Middelgrunden, a huge wind farm in very shallow water two miles outside Copenhagen harbour. Consisting of twenty 2-megawatt turbines arranged in an elegant ellipse, the farm is half owned by an electricity utility and half by over 8,000 individual shareholders, making it the largest mutually owned wind farm in the world. The farm was constructed in 2000, contributes about 3 per cent of the electricity needs of Copenhagen and offers its individual investors competitive financial returns. According to one study, Middelgrunden delivers power for less than 5 euro cents per kilowatt-hour, which is certainly no more expensive than electricity generated from gas.

Denmark and its neighbour Germany have both demonstrated the usefulness of developing community support for wind turbines. by encouraging small investors to participate in the investment and then earning financial returns. In places like Portugal, commercial developers have also provided funding for improved public facilities for the local area. Surprisingly, this model has been slow to catch on elsewhere. Although individual investors or public authorities demand lower financial returns than big companies, the growth of mutually owned wind has been slow outside northern Europe.

Denmark's early support for wind had several important repercussions. Among other benefits, the country became the world leader in the manufacture of turbines. Two of the world's largest manufacturers are based in the country and Suzlon, an Indian company, runs its international marketing from Aarhus in the Jutland peninsula. The other large manufacturers are from Spain, Germany and the US, all countries that have successfully encouraged the growth of wind power, confirming the connection between the expensive process of backing a new technology and the benefit of building a successful industry that can then export to the rest of the world. Countries thinking about investing in other nascent low-carbon technologies should bear this point in mind. Denmark's backing for wind has helped create and sustain a highly valuable manufacturing industry, employing hundreds of thousands of people in good jobs. Eventually, manufacturing leadership will pass to countries with lower labour costs than expensive Denmark, but the country will still capture many of the benefits from its pioneering role.

Suzlon, one of the fastest growing participants in the industry, already manufactures its turbines in India, but almost all of its very substantial research and development work is carried out in Europe. Its lower manufacturing cost base will help it grow its world market share from today's level of about 9 per cent. In its last financial year, its sales almost doubled compared to a worldwide increase of about 40 per cent. Suzlon is particularly well placed to supply China, with its almost insatiable need for power and its wealth of windy locations, such as Inner Mongolia.

Microwind

Even before factoring carbon dioxide into the equation, big wind farms seem competitive across a wide range of locations. For a domestic home a long way away from the electricity distribution network, a turbine on a tower in the garden may provide electricity at a lower price than a diesel generator, especially if oil prices continue to rise. However, for homes and businesses already on the electricity grid, small-scale wind generation will generally not be as financially attractive.

The best way of demonstrating this is to compare the cost per kilowatt of generating capacity. Once the current shortage ends and further technical progress has taken place, a large wind turbine will probably have a total cost of less than $1,000 per kilowatt of maximum power. For comparison, a very small domestic turbine may be priced as much as $6,000 for a kilowatt of peak power, or six times as much. As importantly, today's commercial turbines are elevated many metres above the ground. At these heights, the wind speed is far greater than near the surface. For example, the UK wind speed database says that the average wind speed where I live in Oxford is 4.5 metres per second at roof height but 5.8 metres per second at 45 metres above the ground. The winds would be even higher at 80 metres, the typical elevation of a new commercial turbine. Because the power in the wind is proportional to the cube of the speed, the amount of energy available to collect with a 45-metre mast is over twice what a small rooftop turbine would achieve near my home. Lastly, a tall turbine tower in a commercial wind farm is likely to be located on a site with little wind turbulence. This enables the turbine to capture more wind power *and* reduces the stresses on the equipment, and therefore its need for maintenance.

All in all, for £1,000 invested in a domestic wind turbine, the owner will get less than one tenth of the electricity generated from an investment of the same amount of money in a large wind farm in a windy part of the world. This is not to say that small-scale wind is bad, but merely that for maximum climate change impact, we need to put as much money as possible into turbines that can generate megawatts, not kilowatts.

Not everyone shares this view. Advocates of microgeneration claim, quite correctly, that small wind turbines have the advantage of being visible symbols of a household's commitment to low-carbon electricity. And although the first generation of domestic wind turbines have been widely criticised for not producing as much electricity as their vendors claimed, supporters of microgeneration say that advances in price and efficiency are possible in small wind turbines as much as for their larger cousins.

It is also true that some of the problems associated with turbulent air around buildings and trees may be avoided by wind turbines that do not use the classic three-bladed propeller design. Instead, manufacturers will probably focus in the future on producing what are called 'vertical axis' turbines, such as the one in the photograph. Small microturbines like this will probably be cheaper to build, install and maintain than conventional models of this size.

The machine in the photo, from Mariah Power in the US, costs about $9,000, including installation. In a location with average wind speeds of 12 miles an hour or 5.4 metres per second (breezy, but not exceptionally so) its manufacturers say it will generate 1,800 kilowatt-hours a year – half the electricity needs of a typical European home, though a much smaller percentage of the usage of

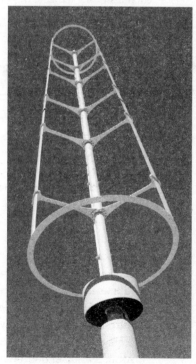

The vertical axis wind turbine produced by Mariah Power

the average US house, which typically uses almost twice as much. The cost of the machine is a major improvement on existing models, but even with generous government rebates it will still take well over a decade before it pays back its owner's investment, compared with as little as four years for a large-scale commercial wind farm.

Technological improvements notwithstanding, cooperatively owned wind farms remain a more exciting proposition than microwind. They encourage support for renewable energy across communities while also offering far greater energy output and financial returns. One US developer, Goodhue Wind in Minnesota, is developing a large 78 MW wind farm with the active financial participation of people in the local area. Many communities around the world could develop a similar strategy of revitalising their district, lessening the cost of power and providing some good 'green' jobs.

Offshore

At the other end of the range of sizes of turbine, what about the viability of wind farms placed offshore? In some countries, but by no means all, onshore wind is unpopular on appearance grounds, which is tending to slow the rate of development. And even in areas that accept, or even actively admire, wind turbines, we will eventually run out of good sites. These difficulties may be avoided in future by deciding to move new wind turbines out to sea, where aesthetic objections are more limited and ecological damage appears to be less severe. Bats, for example, which are threatened by onshore turbines, are very unlikely to fly miles out to sea to find food. So it seems easy to suggest that moving wind generation offshore makes good sense. In April 2009, US Interior Secretary Ken Salazar committed the US government to active development of offshore wind, saying that the coastal resources of almost 2,000 GW exceeded the entire electricity demand of the lower 48 states. The UK government has publicly suggested that 33 gigawatts of offshore turbines might be built off its coasts in the next decade or two, enough to provide almost a fifth of all the country's electricity. Other countries have announced similarly ambitious targets.

However, offshore wind power is in infancy. Although its

potential is vast, the engineering challenges are far more substantial than building even the most remote wind turbines on land. Several significantly sized projects have been completed around the world, but most developers are hesitating before committing to major investments.

The London Array is perhaps the most ambitious scheme for a large-scale offshore farm currently planned. Intending to eventually include over 340 wind turbines, averaging about 3 megawatts in size, the farm will be sited over 12 miles from the Kent coast in southern England in shallow waters at the end of the Thames estuary. Once completed, the Array will produce enough electricity at peak output to power a quarter of the homes in London. A reasonable estimate of the impact of the London Array on the UK's emissions is a reduction of 1.5–2 million tonnes of carbon dioxide per year.

The Array is likely to cost over £3bn (more than $5bn) to build, and even this figure is constantly edging upwards, pushed by the high price of steel and other construction materials. A further problem inflating its costs has been the worldwide shortage of vessels to carry out complicated offshore installations. Another complication is that only two or three manufacturers currently build turbines that can survive years of storms in salty water. Even as the price of turbines reverts to normal levels, offshore-ready turbines are always going to be more expensive than their land-based equivalents.

Because it is offshore – where winds are stronger and more reliable – electricity output per unit of generating capacity will be greater at the London Array than at a typical land-based wind farm. The Cape Wind array of turbines off the Massachusetts coastline should produce even more. The turbines will be rotating a larger fraction of the time. But offshore wind developers need to invest four or five times more, for each unit of generating capacity, than they do for an onshore wind farm. So although offshore turbines produce more electricity, this advantage is outweighed by the huge cost premium imposed by needing to install extremely robust turbines in deep water.

As the industry grows, this incremental cost will erode. Extra offshore construction vessels will be built, more suppliers will enter the market for rugged turbines, and installation techniques will

The Middelgrunden wind farm at the edge of Copenhagen harbour

improve. However, offshore turbines are always likely to be significantly more expensive than their land-based cousins, and building the foundations for the turbine towers will remain a much more difficult task than completing the groundworks onshore. Whether offshore wind power, even coming from huge farms of hundreds of 5-megawatt turbines on 100-metre towers, will ever become cheaper to produce than electricity from conventional power stations is an open question.

Some wind developers are very aware of this problem and have begun to work on designing turbines that float on rafts anchored to the seabed. This approach will avoid the huge costs of constructing underwater foundations and will also mean that the windmills can be installed much further out to sea so that they are completely invisible from the shore. These turbines will never be as powerful as the biggest seabed-mounted models, but the lower installation cost may well outweigh this disadvantage. Statoil, the Norwegian oil company, proposes to place a 2.3-megawatt turbine on a platform similar to an offshore oil-loading buoy 10 kilometres from the Norwegian cost. The company stresses this installation is an early trial, but if it works it may help significantly reduce the costs of offshore wind power.

One might think that offshore wind farms avoid most of the planning permission issues that impede onshore construction in many parts of the world. However, getting the planning permissions and agreements to connect the London Array to the national electricity grid has been a complex and unrewarding task. The onshore sub-station that will receive the electricity as it arrives on land caused a particularly fierce battle. Although the coastline in the part of Kent that will host the substation is not of any great beauty, the struggle to obtain the necessary permissions to construct the building and its outdoor apparatus was delayed for well over a year by inquiries and consultations. Local people were particularly concerned by the prospect of large numbers of heavy lorries passing a primary school on their way to the site during the construction period. The opponents of the substation said there would be thirty lorries a day using local roads; the developer of the Array insisted that the figure was two. This apparently minor debate was one of the two or three issues that significantly delayed the UK's most important new source of renewable energy. Similar arguments were heard over the landing point of the Cape Wind scheme at Barnstaple on Cape Cod. These examples illustrate the scale of the planning problem faced by the wind industry: the London Array, a project that will save carbon dioxide emissions of up to 2 million tonnes, worth about $30m per year in today's carbon market, can be delayed for years by issues such as the routing of a relatively small number of heavy lorries. To many people, clearly, the climate change threat does not appear sufficiently imminent or severe to offset any adverse affects on their local community.

Unreliability – real and imagined

Large onshore and offshore wind farms are going to provide increasing amounts of power over the next decades. Wind may eventually provide 20 or 25 per cent of the total electricity requirements of many large countries. Nevertheless, wind has its problems. Though some people delight in the smooth elegance of rows of three-bladed turbines on the horizon, others think that they despoil the landscape. These opponents usually first attack wind power for its perceived ugliness, but in the next breath they also berate it for 'intermittency'. To the opponents, the key question is this: how can a source of electricity that is so unreliable really be worth investing in?

First, we need to be clear that wind is only really unreliable in one particular way. If the wind is blowing hard now, it will probably also be very windy in a few minutes' time. Very little unreliability there. And, year on year, wind turbines will produce approximately the same amount of electricity over a 12-month period. We have good years and bad years for wind, but annual electricity output from a turbine will stay within well-understood bounds. In that respect, wind turbines are at least as reliable as an old coal-fired or nuclear station, where output can vary enormously because of maintenance needs or equipment failure. Wind is indeed 'intermittent', then, but only in the sense that some weeks it blows hard and in other weeks it doesn't. And even these variations are more predictable than you might think.

So there are three types of intermittency, and wind power only really matches one of these definitions – week-to-week unpredictability but not minute-to-minute or year-to-year. But let's look first of all at minute-to-minute variability. From the viewpoint of the operators of the electricity grid, very short-term reliability matters most. If the total output from all wind turbines on the system did suddenly drop, it would be a significant problem because a well-functioning grid must match electricity supply and demand very precisely. But the wind speed at a single location in the next few hours is actually predictable to a reasonable level of accuracy. And, perhaps more importantly, changes in wind speed tend to be smooth. The wind very rarely drops erratically, unpredictably or quickly although this could, in theory, happen.

Wind's short-term predictability means that this power source is much more easily accommodated in a nationwide electricity grid than might be first imagined. Despite what anti-wind-power campaigners sometimes claim, the people who run our electricity systems do *not* need to keep an equal amount of coal-fired power generation ticking over just in case the wind suddenly drops.

They need a small reserve (perhaps 15 per cent of the total wind-generating capacity) available at short notice, but this is little more than they have anyway. Any national electricity distribution system already has to have power stations ready to start generating electricity at very short notice in case a large power station suddenly fails. Moderate amounts of wind, perhaps up to a sixth of total electricity production, can usually be managed without any changes in the complex way that electricity trading and distribution systems work.

As the total number of wind turbines increases, short-term variability actually becomes easier to handle. Typically, the turbines will be spread over a wider area – perhaps the whole country – and when the wind is quiet in one place, it is likely to be blowing strongly in another. The total electricity output from 1,000 turbines varies far less than the power generation from ten.

But there is a limit to the amount of wind power a grid can easily accommodate. In some short periods during March and April 2008, wind turbines in Spain were generating up to 40 per cent of the country's electricity requirement. At this peak, the grid operator obliged electricity generators to disconnect some turbines from the network. Although the wind levels were extremely unlikely to drop unexpectedly, even small percentage changes in wind output might have overwhelmed the country's limited ability to import power instantaneously from France or switch on back-up power stations. This is a genuine concern arising from the growth of wind power – the tiny but important risk of unpredicted short-term variability when wind is responsible for a very large percentage of electricity generation.

The solutions to this problem are reasonably simple, although not necessarily cheap. We could simply build cheap gas-fired power stations that can be very quickly fired up in the event of unexpected shortages of supply – though this means cost and carbon emissions. Alternatively, we could try different approaches that do not require us to burn more fossil fuels whenever wind speeds don't match expectations. We have three main routes for achieving this. First, we can make it easier to import power from remote locations. Second, we can 'store' electricity. Third, we can introduce systems to manage electricity demand at short notice so that it matches the available supply.

Every country in the world that relies on increasing amounts of wind, marine or solar power will probably need to use all three of these mechanisms for aligning short-term supply and demand. In the US, this three-pronged approach is appropriately called the 'smart grid'. The building and operation of this new kind of grid is a fascinating challenge to engineers and also the the mathematicians who will use statistical modelling to minimise the risk of not having enough power or, perhaps even more expensively, having grossly excessive power production for many hours a week.

Elsewhere, the standard approach, which we might call the 'twentieth-century model', simply tries to predict changes in demand and then adjusts supply to meet these variations. For example, an advertising break in a popular television programme produces a sudden surge in demand as lights are turned on and kettles boiled. Just before this happens, a large power station is warmed up, ready to start producing electricity the moment demand begins to surge. Supply simply aims to match moment-to-moment demand. This model is both costly and carbon intensive, because power stations have to be held in reserve, burning large amounts of fuel even when they are not supplying power to the grid.

The smart grid is more efficient – and it's also compatible with the incorporation of large amounts of power from wind and other unreliable energy sources. Let's look in a little more detail at its three main approaches.

Importing remote power

If the wind suddenly drops on the Atlantic coast of Spain, it is statistically extremely improbable that Denmark will suffer at the same time. So if the electricity grid connected Spain and Denmark, spare power could flow southwards at very short notice. At the moment, most countries have poor connections to their neighbours. When Spanish wind generation peaked in spring 2008, the link between France and its southern neighbour was not robust enough to handle the possible demand from Spain if the wind suddenly dropped. It can currently handle only about 5 per cent of

Spain's total demand. Growth in wind capacity around the world must be accompanied by major investments in power distribution networks. This means increasing the number and size of the electricity transmission links between different countries and between regions inside countries. The aim is to be able to move electricity nearly instantaneously from countries in surplus to those in deficit. These new high-voltage links will need to be paid for, but most independent studies show that the costs are unlikely to add more than a fraction of a penny to the price of power. Building bigger, better and more robust electricity grids simply must happen if we are to significantly increase renewable power use. Most governments understand this but many still do not.

Storing electricity to meet short-term needs

We can store electricity in batteries, but this is expensive. It does not provide a large reserve and cannot yet be used to back up wind turbines. Today, the cheapest way of providing a reserve of usable power is through the use of a system called 'pumped storage'. When electricity is abundant or cheap, it is employed to move very large quantities of water uphill into a storage reservoir. At moments when demand is very high, or when the grid needs power urgently, the water in the high reservoir is released, turning turbines as it falls back into the lower reservoir. A good pumped storage system can start generating large amounts of power as little as 15 seconds after a request from the electricity grid. The UK's largest storage reservoir, in north Wales, can supply as much power as the biggest power station in the country and is invaluable on the infrequent occasions when a large power plant does fail without warning. When the higher reservoir is full, it can keep generating for several hours before it runs out of water. This gives the people who run the electricity network enough time to start other power stations.

For many countries, pumped storage is the best-established way of dealing with immediate needs for power. The Spanish electricity grid now has about 3 gigawatts of pumped storage. As in the UK, this capacity was built to help insulate the power grid from the effects of spikes in demand or major power station failures. But now, in addition, the country has over 15 gigawatts of wind power, expected to rise to more than 20 gigawatts by 2010. The ratio between the current storage capacity and the possible fluctuations of wind output is not great enough. As a larger and larger percentage of electricity comes from renewable sources, the need for countries like Spain to build water storage reservoirs will grow. This is not necessarily easily achieved. To provide a satisfactory site, the operator needs to use two large reservoirs, not far apart, one much higher than the other. Many countries will have relatively few locations that meet these characteristics. Moreover, those places that are suitable will tend to be in hilly or mountainous places far away from the main electricity pylon links, making it difficult to connect the reservoirs to the distribution grid.

One alternative is to establish lagoons at sea – large areas of ocean surrounded by a high wall. When electricity is plentiful, pumps bring seawater into the lagoon, creating a gradient. If the wind falls unexpectedly, this can be used to drive turbines for near-immediate electricity. These barrages may cost no more than building pumped storage sites in the mountains, but no country has yet invested significant amounts of money in developing this approach.

Other ways of holding a reserve of electricity include making hydrogen – by splitting water into its constituent atoms – or compressing air. Using electrolysis to separate the hydrogen and oxygen atoms in water at time of abundant electricity, storing the hydrogen, and then burning it to drive a generator when electricity is in short supply is a plausible alternative to pumped storage. Similarly, we could use surplus power to compress large quantities of air, store it in depleted oil or gas reservoirs or salt mines, and then utilise the power of the expansion of the air to drive generators when the power is needed. Both of these techniques look more expensive than water storage, though little work has yet been done to firmly quantify the costs. Pumped storage is also probably more efficient than these alternatives in the sense that a greater fraction of the stored energy is recreated as electricity when it is needed. The downward flow of the water can recapture about 70 per cent of the energy needed to pump the water upwards in the first place. Compressed air and hydrogen manufacture don't look as though they can generate quite the same return, although the differences are not large. In August

2009, Pacific Gas and Electricity announced a plan to build a trial compressed-air storage facility that could provide as much electricity as a medium-sized power plant for about 10 hours.

Managing demand to meet supply

Supply and demand need to balance almost exactly on an electricity grid. Otherwise the voltage or frequency of the alternating current would move outside the tolerances of home appliances and business equipment, possibly causing damage. Pumped storage provides a way of rapidly adjusting supply, but it's also possible to almost instantaneously reduce demand. In the jargon of the electricity industry this activity is known as 'load shedding'. Some manufacturing companies, for example, have agreements that allow their electricity supplier to disconnect them at a few moments' notice. In return, the companies are given lower prices. This system works well. Though it is designed primarily to shave a little off the sharp daily peaks of electricity demand, there is no reason why the same approach couldn't be adapted to deal with temporary shortfalls in wind power at all times of day.

In the US, some electricity companies operate a slightly different scheme. They pay customers a rebate every month for promising to immediately reduce their electricity consumption when asked. For big commercial customers this might work out as a $100 per year saving for every kilowatt of committed reduction. So if a company usually uses 400 kilowatts to power its office block, but agrees to reduce this to 100 kilowatts at a few seconds' notice, it would be paid $30,000 a year. These are voluntary programmes but they work well because most users can cut their demand easily and at minor inconvenience. Shops can reduce lighting use, hospitals can turn on their emergency generators, and businesses can shut off their air-conditioning for an hour or so. To the electricity provider, these reductions may be equivalent to having an extra power station available – though a spare power station, sitting ideal for 90 per cent of the time, would be a much more expensive and polluting solution.

The major load-shedding programmes around the world typically cover about 2 per cent of peak power use. This is enough to cope with temporary energy deficiencies as long as wind is not too great a percentage of total electricity supply. For Spain and Denmark, countries that occasionally have wind providing a very large fraction of total electricity demand for several hours at a time, this scheme would not be enough on its own. However, many people in the electricity industry think that it ought to be possible to increase the percentage to 10 per cent or even more. This would be enough to protect the power grid from wind's small minute-to-minute variability in almost all circumstances.

Individual households can also be encouraged to reduce demand on a signal from the national grid. More and more countries intend to provide homes with 'smart' electricity meters that can be remotely instructed to switch appliances off or which can limit total household power use to a set level – say, 3 kilowatts. Already some French homes are fitted with meters that restrict energy consumption to this level. In Italy almost all the customers of the main electricity company have smart meters and can reduce their bills by switching their use of electricity to the periods of the day when prices are cheapest. More advanced meters could be used to switch off non-critical machines such as dishwashers and washing machines at moments when wind power drops. The technology is already available to do this. A signal carried over the mobile phone network might trigger an electronic on/off switch at the wall socket of those electric appliances that use large amounts of electricity.

A programme involving all three of these techniques for balancing supply and demand – better electricity grids, greater energy storage and load shedding – allows electricity operators to deal with the occasional unexpected drop in wind power. There will be some costs, and some inconvenience, such as dishwashers and washing machines suddenly switching off for an hour or so, but it can be done.

In many parts of the world, regional electricity grids are already finding ways of reducing peak demand. In southern parts of the US, the electricity suppliers often struggle to cope with summer afternoon peaks resulting from the increased use of air-conditioning. The arrival of large amounts of wind energy means that the problem of matching supply and demand is becoming more urgent: the grid has slightly more variability of supply in addition to fluctuating demand that can rise very rapidly as afternoon temperatures increase. In the most vulnerable locations, electricity companies are particularly interested in finding ways of increasing the amount of load shedding available to them.

Other techniques for managing supply and demand will become available. As Chapter 6 explains, the batteries in electric cars could also form a vital buffer to keep the electricity system stable. When enough people own electric cars, their batteries will offer an extremely attractive alternative to other ways of matching short-term fluctuations in supply or demand. Four million car batteries, each providing 3 kilowatts of power, would match the maximum output ever achieved by all Spanish wind farms. Because the batteries of electric cars contain so much energy, this could continue for several hours before it had much impact on the state of charge of the vehicle fleet. Intelligent electronics, connected to the grid, could detect when power was needed and instruct parked, plugged-in cars to start supplying power. Conversely, if the grid was oversupplied with wind, the batteries could soak up the excess. When we have enough of them on the road, electric cars will have enough capacity to keep the whole electricity system stable. Proponents of renewable energy sources such as wind and solar should therefore also be committed enthusiasts for electric cars.

These measures can deal with the relatively small problem of very short-term and unexpected fluctuations in wind output. The absence of productive levels of wind for long periods is a more difficult problem, even when this absence has been predicted many weeks in advance and a country has access to large amounts of power from other regions. The possibility of days and weeks of low wind speeds represents a real challenge to the operator of the electricity grid. One response is to have a very large amount of unused generating capacity, probably using gas as a fuel source. These plants won't need to be ready without warning, but need to be ready to start generating with perhaps a day or a week's notice. Some grids are already investing in relatively inexpensive gas-powered turbines to provide this back-up power, but there are other routes forward.

In most of Europe, the availability of wind tends to be greatest when electricity demand is highest: in the early evening of the winter months. That's partly because of the simple fact that there's more wind in winter, when European electricity demand is at its peak. However, it's also because windy conditions in themselves tend to add to electricity demand. High winds increase the heat loss from houses. Those houses heated by electric appliances will usually need more electricity when wind energy is most available. In these parts of the world, the supply of wind-generated electricity is strongly correlated with the demand for power.

In some hotter countries this happy relationship of supply and demand does not occur, because summer air-conditioning is often needed exactly when the wind is not blowing. In these countries, the obvious climate-friendly solution is to install large amounts of complementary solar power to meet summer peak demand.

In fact, solar energy is likely to be the best way of balancing wind supply in most countries, because of the inverse relationship between the amount of wind energy available and the strength of the sun. For the countries of Europe, a logical mixture of power generation would see concentrated solar power from North Africa (see Chapter 2) providing much of the summer electricity, with wind and wave power – both most productive in winter – taking most of the strain in colder months. Tidal energy from geographically dispersed sites, generating maximum power at different times of the day, could provide a solid base of power availability, particularly at the equinoxes. In reserve, and ready to fire up if adequate supply looked uncertain, could be large numbers of wood-burning power stations. Unlike other forms of renewable generation, wood power plants need to pay for their fuel, but the cost of the generating equipment is not particularly high. It therefore makes sense to use wood as the main back up for when all other renewable sources cannot match demand over periods of days and weeks.

We will still need gas and coal stations. So even countries with extraordinary resources of renewable energy, such as Scotland, with its wind, waves and tides, or Spain, with its sun and wind, will still need to work on capturing and storing the carbon dioxide emitted from the remaining fossil fuel power generation, as discussed in Chapter 8.

More wind myths

The opponents of wind energy focus not only on the perceived ugliness of turbines and the unreliability of the power. Criticism is also directed at the potential impact on wildlife. Many of these concerns are unwarranted and others can be exaggerated. Most land animals get used to turbines very quickly. Horses and cows, for example, ignore the rotating blades very soon after they are installed. Local birds are also largely unaffected by wind farms, although one wind farm in Norway has probably been responsible for almost wiping out a colony of rare eagles. Migrating birds may be more of a problem, though the effect is still utterly insignificant compared with, for example, the impact of casual hunting in Italy and France, or the numbers of birds killed by road traffic or predatory cats. More troublesome may be the impact on bats, which appear to be poor at avoiding the moving turbine blades.

Another frequently repeated criticism of wind power is the suggestion that the energy embedded in the manufacture and installation of a turbine is so great that it counterbalances the greenhouse gas reductions from several years of operation. This is simply not true. Research invariably suggests that wind turbines pay back the energy invested in them within a few months. Of course, it depends on the windiness of the site and on the amount of concrete that needs to be used to create foundations and access roads, but commercial wind power seems to have a strongly advantageous payback of energy and carbon dioxide. (There is one possible exception to this conclusion: wind farms constructed in areas of thick peaty soils may result in substantial emissions of globe-warming methane and CO_2 from the drying out and rotting of the peat. Good construction techniques that avoid undue disturbance of the surrounding landscape can avoid this potentially severe problem, however.)

One way of checking this favourable conclusion is to estimate the carbon dioxide saved by using wind turbines and compare this with the emissions produced in making the steel for the turbine. We can do this very roughly by looking at the figures for Adam Twine's wind farm.

Each one of Adam's turbines will generate about 2.5 gigawatt-hours per year – enough to provide the electricity for perhaps

600 local homes. If the wind farm hadn't been constructed, this electricity might instead have been generated at Didcot coal-fired power station, ten miles away. Didcot would have emitted at least 2,000 tonnes of carbon dioxide from burning the coal necessary to produce this amount of electricity. One medium-sized wind turbine therefore saves these emissions every year.

How does this compare with the energy needed to make the turbine and its tower? Over 90 per cent of the weight of a typical turbine is steel and a large fraction of the total energy used to manufacture this steel arises in the smelting of the metal, usually in a blast furnace. An efficient modern steelworks emits about 2 tonnes of carbon dioxide per tonne of finished steel. The weight of each of the turbines on Adam Twine's farm, including the blades and the supporting steel pole, is less than 200 tonnes. So the emissions from making the steel will be no more than about 400 tonnes of carbon dioxide. The cement in the concrete foundation and the ground works to allow access to the wind farm will have added to this figure, perhaps bringing it up to 500 or 600 tonnes. Compare this with the 2,000 tonnes of emissions that each turbine will save *every year*. By these figures, the greenhouse gases arising from producing and installing the turbines will have been outweighed by the savings in emissions at Didcot within just four months of use. This informal calculation is not precise, of course, but it demonstrates that a wind turbine is likely to produce perhaps a hundred times more energy, in its lifetime, than is used in its manufacture.

One final concern occasionally raised about wind power is that the erection of thousands of turbines might radically change local or global weather patterns by slowing down the speed of the air. This worry might have validity if turbines captured more than an infinitesimal share of the total energy in the moving wind around the world. Any significant change in global weather patterns would probably only occur if a measurable fraction of the world's surface was devoted to wind farms. Today, the reduction of wind speeds as a result of new turbine construction is almost certainly less than the increase in wind levels caused by the world's loss of forested area. Trees slow down the wind, too.

Solar energy

Enough to power the world many times over

The sunlight hitting the earth's surface every day contains around 7,000 times more energy than the fossil fuels that humanity consumes. If we could find an economical way of exploiting this energy, then all the world's energy and emissions problems would be solved. Even with today's technologies, solar collectors on less than 1 per cent of the world's unused land could comfortably match all fossil fuels. Which is no surprise when you consider that a sun-soaked tropical area of just 10 square metres – approximately the floor space of a small bedroom – receives as much energy from the sun as the typical global citizen consumes for transport, heating, food, electricity, and all other aspects of life.

Of course it isn't as simple as that. Aside from anything else, the sun's rays are at their most powerful in the tropics, while much of the world's population is found in temperate countries thousands of miles away. But the potential is huge, and solar technologies have many advantages. Not only are they climate friendly, they're also non-polluting, almost noiseless, and require little maintenance. In addition, unlike biomass energy, they make use of non-productive space – be it deserts or urban rooftops – and therefore don't put pressure on food production.

There are three main ways of capturing the sun's energy. The first is to put long tubes containing liquids in direct sunlight. The liquid in the tubes gets hot and, using a heat exchanger, can be used to heat water for showers or for washing clothes. The second way is to use panels of photovoltaic (PV) cells to turn the photons of light directly into electricity. Finally, there are solar concentrators, which use mirrors to focus large amounts of sunlight on to a small area, intensively heating fluids and then using their energy to drive a turbine or a Stirling engine to generate electricity.

The first of these approaches – solar water heating – has been available for centuries. It is a straightforward technology and can be remarkably efficient. On our house in cloudy Oxford we have forty glass tubes about 2 metres long on the roof. Inside the tubes is a thin flat foil of copper. This foil is heated by the sun and transfers the heat to a liquid in a thin pipe running in the centre of the tube. This liquid is pumped into a heat exchanger that transfers the energy to the hot-water tank. It is an extremely simple and reliable system. On a summer's day in southern England, it provides all the hot water for five people. It has captured perhaps 70 per cent of the light and infrared energy falling on the glass tubes and transmuted it to useful hot water. No other solar technologies are anywhere near as efficient. But heating water for baths and showers is not a large part of the energy needs of most households. In Europe, heating hot water needs less than 1,000 kilowatt-hours per person each year, or less than 3 per cent of total energy requirements. Nevertheless, in sunny countries, solar hot water makes good financial and environmental sense. Cheap solar collectors can provide heat for hot water for most of the year. In high-latitude countries such as Britain, I have to admit that solar water heating will barely cover its cost, even at today's fossil fuel prices.

If we are to get the greatest impact from solar energy, we should use it to generate electricity, not hot water. Indeed, we need to get solar electricity to the point where it offers developing economies a cheaper way of fulfilling their growing needs for power than from burning fossil fuels. Higher-latitude countries may find wind energy the cheapest form of power, but solar power will eventually be the best way of generating electricity in the tropics, where wind speeds are generally lower.

The traditional way to produce electricity from sunlight, by means of photovoltaic cells, is well established but still expensive. This chapter explores the various ways that scientists and technologists around the world are trying to bring down costs and raise efficiency. Some of their innovations are very promising. However, the chapter also argues that mirror-based solar concentrators are just as significant and offer huge potential. Prince Hassan bin Talal of Jordan, a leading backer of this technology, outlines the vision:

> In deserts, clean power can be produced by solar thermal power plants in a truly sustainable way and at any volume of conceivable demand. This gives the deserts a new role: Together with the many other forms of accessible renewable energy the newly utilized desert would enable us to replace fossil fuels and thus end the ongoing destruction of our natural living conditions.

The idea of using the sun of the Middle East and Africa to provide Europe with limitless and cost-competitive power is hugely appealing. Networks of enthusiasts for this project have sprung up all around the region, led by scientists and electric power utilities in Germany. Solar power concentrators have immense potential around the world, not just in Europe. The arid south-western lands of the USA are particularly suitable but any area of sunny desert will provide huge quantities of energy, potentially at relatively low cost. As we'll see later in this chapter, the challenges are primarily logistical and commercial, not technological.

Solar photovoltaics

In 1958, the US launched Vanguard, the first satellite equipped with solar photovoltaic cells to provide electric power. The panels produced about 1 watt of electricity, not enough to cover the passive use of electricity in an idle TV set today. Only a minuscule fraction of the energy hitting Vanguard's solar panels was turned into electric current. Half a century later, and advances have taken the efficiency of some commercial photovoltaic panels to a maximum of just over 20 per cent, though the cost remains oppressively high.

Simply put, a photovoltaic panel creates electricity when light energy (a photon) hits the silicon surface and pushes an electron out of the top layer of the silicon and across an electrical junction inside the panel. The movement of this electron creates a useful voltage. When wires are connected to produce a circuit, this voltage means that current will flow, eventually taking the displaced electron back to the top layer. Solar cells work best in strong sunlight, but will also generate some power on an overcast day from the diffused light that gets through the clouds. They're not particularly complex devices, but the technical challenges in producing them cheaply are formidable.

Most solar panels manufactured today are made from expensive slabs of extremely pure silicon. The silicon is derived from a very abundant substance, common sand, but the process of refining it and ensuring that it is pure enough for electrical use (and for the semiconductors used in electrical devices) is complex and energy intensive. As the solar panel industry grew, encouraged by enormous subsidies in Germany and other countries, the supply of pure silicon did not keep pace: 2008 saw significant shortages that pushed the price of solar panels up. The price rise has been followed by a sharp decline, however, as large numbers of new factories in China and elsewhere began producing unprecedented volumes of silicon in 2009 and as some of the major markets, such as Spain, saw sharp reductions in the financial incentives to install PV systems. By the last quarter of 2009, the prices of PV modules were more than 20 per cent below the levels of mid-2008.

Solar panels can be built in any size. Small rectangles of silicon provide the power for personal calculators and other minor domestic appliances. Much bigger blocks are used to make conventionally sized solar panels, which are over 1 metre tall and somewhat less than 1 metre wide. In a solar power station, huge numbers of these panels can be chained together, all providing electricity and perhaps generating as much as 50 megawatts in full sun. As solar panels decline in price, which should happen rapidly over the next ten years, we can expect to see solar panels installed in larger and larger groups, with total power output close to that of conventional power stations.

By early 2009, the installed photovoltaic panels across the globe could produce about 15 gigawatts of electricity if working in full sun, somewhat less than 10 per cent of the worldwide capacity of wind power. (The power of a solar panel is usually expressed as the maximum output when the sun is showing strongly at midday.)

Most of the time, the electricity actually produced will be much less. In a very sunny country, a day's electricity from a panel with a rated power of 1 kilowatt might be about 6 kilowatt-hours. This is about a quarter of the electricity that would be generated if the panel were working to peak efficiency for the full 24-hour period. This typical performance implies that all the solar photovoltaic panels in the world currently provide less than 10 per cent of the total electricity demand of a country the size of France or little more than 10 per cent of the needs of the state of California.

Solar electricity is growing rapidly, perhaps by 30 or 40 per cent a year, but today's global photovoltaic output is only equivalent to a couple of very large coal-fired power stations or a small cluster of nuclear plants. This comparison helps show the enormous scale of the challenge. After fifty years of research and development into photovoltaic technology, we are still only obtaining a small fraction of 1 percent of world electricity demand from solar sources. So why should we be optimistic that all this will change and that solar energy will provide a significant percentage of world electricity within a decade or so?

The cost of photovoltaics

The financial performance of solar panels has consistently improved over the years as technical progress has reduced costs and raised the output of electricity. But PV is still a very expensive way of generating electricity except in the sunniest places.

If installed today, the large and slightly ungainly slabs of silicon on the roof of my home and the associated electronics attached to the wall of the garage would cost at least £12,000. In roughly four years on a house in Oxford, they have produced about 6,000 kilowatt-hours of electricity. If I'd sold that power in the wholesale market for electricity, I would have banked a cheque for about £300. The panels will probably last another twenty-five years before their performance begins to degrade, so the total value of their power output without subsidy will almost certainly never cover the original cost. In our case, a government-sponsored capital grant available at the time of installation (but not now) and the enhanced prices we obtain for the electricity we export means that the panels will actually earn our household a reasonable return of at least 5 per cent a year on the capital we invested.

On the roof of a house in high latitudes, photovoltaics don't seem a cheap way of reducing carbon emissions. Partly this is because putting panels on a tall house requires expensive scaffolding and several days' work. It's also because the two 'inverters' needed to change the low-voltage direct current output from the panels into the 240 volts alternating current required by the electricity system contain sophisticated electronics. Each inverter cost £1,000 four years ago, and the total amount of power we have generated thus far wouldn't even cover the cost of these devices.

Germany introduced generous 'feed-in tariffs' in 2000, aimed at encouraging faster installation of renewable energy. Property owners who put solar panels on their roofs are entitled to substantial payments for every kilowatt-hour of electricity that they feed into the local electricity grid. The incentives were, and still are, enormous. Each kilowatt-hour of solar energy is currently worth about 45 euro cents, or six times the typical wholesale price for electricity. The feed-in tariff set the solar photovoltaic industry alight in Germany and, by the end of 2007, over 300,000 homes and businesses had solar roofs. California, by comparison, had only about one tenth of this number in early 2009. The cost in Germany has been enormous – probably more than $7bn a year for less than 1 per cent of its total electricity need – but the subsidy has built expertise and knowledge. It is little exaggeration to say that the world solar power industry as we know it today would barely exist without the subsidy to German householders.

The International Energy Agency criticised the expense in a 2007 review. It said:

> Estimates show that between 2000 and 2012, the feed-in tariff will cost €68 billion in total. In particular, the subsidies provided to solar photovoltaics are very high in relation to output; they will eat up 20 per cent of the [renewables] budget but contribute less than 5 per cent of the resulting generation.

High payments to the owners of solar panels may have cost other

German householders large amounts of money, but there is little doubt of the effectiveness of the policy at helping photovoltaic manufacturers across the globe reduce their costs and improve their production processes. At one stage over half the solar electricity produced in the world came from German roofs, even though the average German solar panel produces substantially less than half as much electricity as the same panel would in the Sahara desert. Those of us not living in Germany should be grateful for the generous solar subsidy and in its impact on manufacturing costs down across the world.

The billions of euros spent overpaying Bavarian farmers to put solar panels on their cowsheds have attracted widespread political attention and other countries have copied the idea of feed-in tariffs. France and South Korea, to give just two examples, are following the German lead. But no country wants to subsidise renewable energies forever, and panel manufacturers around the world believe that solar photovoltaic technology must rapidly become competitive with fossil fuels if it is to continue to prosper.

As a result, all the competing companies vying to build big businesses in solar technologies have one target in mind. For solar to become truly competitive with fossil fuels, these businesses say that photovoltaic panels have to cost the customer no more than about $1 per watt of maximum power, or about $1,000 per kilowatt of peak power. The costs of the associated electronics and the expense of the installation will probably double this price, taking up it to $2,000 a kilowatt. This is about a fifth of the cost of the equipment installed on the roof of our house four years ago, so the challenge is enormous.

Of course, solar doesn't generate much electricity when the sun isn't shining, so the 'dollar per watt' figure isn't easy to compare with the cost of a coal plant or natural gas power station. In a sunny region, solar panels with a capacity of 1 kilowatt will generate over 2,000 kilowatt-hours a year. Wholesale electricity prices vary around the world but this typically might be worth $200 to $250 to the power producer (less in the US and some other markets). Since solar energy has very low yearly operating costs, the payback period on the initial investment of $2,000 per kilowatt might be as little as eight or ten

A large installation of photovoltaic panels

years in countries with expensive electric power. If solar photovoltaic technology is any more expensive than this, it will require continuing subsidy. The big US, Chinese and Japanese panel manufacturers are only too aware that the overgenerous 'feed-in' prices paid for solar energy exported to the national electricity grids in places such as Germany are already edging downwards. Electorates will not be willing to support high-price solar electricity indefinitely. Very sensibly, these manufacturers are therefore aiming to get their production costs down to a level that implies that customers will not need subsidies to justify buying photovoltaic panels.

The 'dollar a watt' target is also known as 'grid parity', that is, the point at which the industry believes it is no more costly to use solar power than fossil fuel power plants to supply the electricity distribution grid. Progress is surprisingly fast towards this objective. One recent industry study said that solar technologies will be competitive with coal and gas by 2015 across most of the US, excluding only the least sunny areas.

The largest US solar panel manufacturer, First Solar, is even more optimistic. It expects to achieve grid parity no later than 2012 and several other companies, some using very different technologies to make their panels, have made similar claims. By 2012, First Solar will probably be supplying several gigawatts of PV panels each year to its customers, largely to companies building huge farms of solar panels in sunny areas. Photovoltaic technologies tend to work less well in high temperatures, so these solar power stations will tend to be located in areas with good solar radiation, but which are relatively cool. South-facing mountain slopes are ideal.

Having to drive down costs 20 or 30 per cent a year is one problem for the relatively small number of large panel manufacturers. The other issue they face is even more challenging. There are at least four different types of solar panel, using different semiconductors in a variety of thicknesses. The old technology – heavy layers of pure silicon – is under threat from upstart new approaches, often using very thin coatings of semiconductors on a simple backing material. In this exciting but still obscure industry one of the world's great business battles is just beginning. Not entirely friendly controversy rages between manufacturers over which method will ultimately prove to be the best way to compete with fossil fuels. Most of the companies building their advanced new factories have committed irrevocably to one technology or another. Many billions of dollars ride on success.

Which of these options will give us electricity at the lowest price per kilowatt-hour? And which will give us the fastest rate of increase in the total generating capacity installed around the globe? 'Old-fashioned' panels, manufactured from large amounts of silicon, have the advantage of capturing a relatively large percentage of the energy of the sun but are expensive to manufacture. Is it better to focus on much thinner films of silicon that deliver smaller amounts of electricity from every panel but at a substantially lower initial cost?

The next question is whether the best material is silicon at all. First Solar has opted for a semiconductor called cadmium telluride, from which it makes thin panels that are relatively inexpensive but only moderately efficient at capturing the sun's energy. First Solar's cells convert about 10 per cent of the sun's energy into electricity, though the company is planning improvements which will take this up to 12 per cent or more within a few years. Other companies, including the secretive Nanosolar, funded in part by the billionaire founders of Google, are concentrating on another semiconductor material, known as CIGS (copper indium gallium (di)selenide). Nanosolar's backers are hoping that its revolutionary technology, which simply 'prints' the semiconductor material on to a flexible metallic backing layer, will prove an approach that offers panels of such low cost that they will be wrapped around the exterior of millions of buildings across the world. As its name indicates, Nanosolar is using nanotechnology to precisely arrange the atoms on the printed semiconductor surface. Its frequent claims that this approach will eventually produce extremely cheap panels are convincing to many outsiders but treated with undisguised scepticism by other businesses in the PV industry. After several years of R&D, the reclusive company finally shipped its first commercial panels to a solar farm in Germany in the last days of 2007. In September 2009, it opened its first factory in Germany, able to continuously produce solar panels at a rate of one every few seconds. If Nanosolar's most important boast – that it can print solar PV cells of one hundredth the thickness of conventional silicon panels at speeds one hundred times as fast as current manufacturing processes – is even partly true, the cost of PV is likely to fall extremely rapidly as it ramps up its production.

The battle is not yet over. Some large manufacturers are sticking with conventional thick silicon solar panels, believing that the price of pure silicon will eventually fall dramatically, meaning that their raw materials will be much less costly in the future and it won't matter if they use several kilograms in each panel. The Japanese company Sharp, which has been making solar panels since 1959 and has a large market share of solar installations around the world, is focusing its efforts on improving the conversion efficiency of its silicon panels rather than reducing the amount of silicon it uses.

The competition between all these alternatives is essentially between those who believe that reducing the cost per square metre of panel is the most important objective and those who think that it is more important to focus on improving the efficiency of capturing solar energy. Companies focusing on thin films, whether of silicon, cadmium telluride or CIGS, are betting that the vital change is getting panel prices down to low levels. If the panel is very cheap to make, then it doesn't matter much if you have to use a larger area.

On the other side of the argument, those concentrating on improving the efficiency of light-to-electric conversion think that panel costs are not the most important element. They claim that the cost of panels, both thin film and conventional, will eventually fall substantially but point out that this cost is only a part of the bill for a solar installation. Eventually, they predict, the panels will represent less than half the total cost of putting a solar roof on a large commercial building, with labour, cabling and inverters accounting for most of the rest. Therefore, the argument goes, the most important thing is getting the largest possible efficiency per area of roof and per dollar of installation costs.

The companies betting that the future of PV is based around common silicon, whether thin film or conventional, attack their competitors on another front. They say that cadmium telluride and CIGS companies both need reliable supplies of chemical elements that are in dangerously short supply. First Solar, the hugely successful business based on cadmium telluride, frequently has to rebut assertions that the world is simply going to run out of tellurium in the next few years. Although only a few grams of tellurium are needed for each panel, the cost of this rare metal has risen dramatically in the last few years. First Solar says it has guaranteed supplies for several years ahead. Other supporters say that tellurium is available in very large quantities on underwater ridges beneath the oceans where it can be cheaply mined.

CIGS is the subject of a similar debate. Sceptics make pessimistic statements about future shortages of the indium metal in CIGS. Here the issue is not so much absolute scarcity: indium is at least as common in the earth's crust as silver is. But the current generation of LCD screens, used for computers and televisions, is competing for the relatively small amounts of the metal currently mined every year, and the price has risen sharply. Nevertheless, indium prices probably won't be a substantial problem for CIGS manufacturers in the future. Within the next ten years video and computer displays are likely to be made using a different technology that does not use the metal. The CIGS manufacturers are gambling that indium will become cheaper as supplies increase and demand from competing products eventually falls away.

There is one further complication. The maximum level of electrical efficiency for standard panels is about 20 per cent. (Thin film panels struggle to reach 10–12 per cent.) This low figure arises because current PV panels are only able to capture the energy from a small portion of the visible light spectrum. Red light passes through the panel without dislodging electrons, while blue light is largely reflected. In theory, a PV device that combined several layers, each with different absorption characteristics, could capture far more energy, and perhaps even exploit invisible infrared light. Panels like this are called 'multi-junction' cells, and they've already been demonstrated in laboratories with efficiencies of nearly 40 per cent.

The ultimate aim must be for the manufacturing companies to make cheap, thin, printed, multi-junction solar cells, probably from inexpensive silicon. When this happens, we will see buildings around the world covered in these panels and generating all the electrical energy that they need at a price to beat any electricity delivered from large, remote power plants. The crucial question is when. The amount of private capital going into PV technology is large enough to achieve the objective of very low panel prices but my suspicion is that mass availability is as much as ten years away. Although all the major photovoltaic manufacturers are publicly saying that advances in cost and performance are going to be extremely rapid, I think a little scepticism is probably justified.

Given the slow pace of progress in PV over the past half century, why should anyone be even this optimistic? First, nanotechnology really does make a difference. Now that companies can make specialised materials whose atoms are very precisely arranged, we are seeing rapid advances in the ability to capture the energy of the photons hitting the panel. Second, the impact of the German feed-in tariff has been to vastly increase the total number of panels being made around the world. The effect on manufacturing costs, ignoring the temporarily very high price of silicon, has been dramatic. The world is only making a few gigawatts of PV panels each year at present, but we are doubling the accumulated manufacturing volumes every couple of years. The cost reductions achieved so far from moving down the learning curve give us good reason to believe that as volumes continue to increase we will see continued very sharp declines in cost.

Eventually, PV will almost certainly be the technology of choice for small-scale and localised generation of electricity in sunny countries. With luck, low-cost solar panels will be available to meet the needs of remote communities in Africa and Latin America well away from the electricity grid. In other words, these places may never need to install fossil fuel power stations. All that's needed is a low-cost storage technology to cover nights and periods of cloudy weather. This bypassing of fossil fuels is already happening on a very small scale. The German company SunTechnics is supplying panels to Namibia, where many of the people live far from a reliable electricity supply. The electricity users do not buy the solar panels and other electronics, but simply pay for the electricity that they use via a prepayment system. The utility company that operates this service is, in effect, renting the solar kit to the household or business and can move it elsewhere if the customer no longer wants the power or turns out to be a bad credit risk.

In developed countries, solar PV will eventually make most sense installed on the user's premises, rather than in the huge centralised power stations that First Solar is currently focusing on. The primary reason for this is the fact that a commercial solar farm feeding into the national grid will get paid the wholesale prices for power, which are typically about 50 per cent of the price paid by homeowners or small commercial customers. By contrast, solar PV installed on homes or offices displaces power that the building user would otherwise have purchased at the retail price. PV is one of the few electricity sources that can be installed on a very small scale and still be reasonably productive. A wind turbine on a house costs ten times as much as a commercial wind farm per unit of electricity generated, but the comparable ratio for solar PV is probably only about two. This means it may eventually make good financial sense for a building owner to put PV on the roof, displacing electricity supplied by a utility. The capital cost disadvantage of a small installation does not outweigh the savings from avoiding having to pay retail prices for power.

The largest problem, as with some of the other technologies in this book, is scaling up the manufacturing of solar panels at a fast enough rate to dent global greenhouse emissions in the short window of time available. There are perhaps twenty or thirty companies in the world currently producing large numbers of advanced PV panels, or which hope to be manufacturing them in the near future. First Solar hopes to be making enough panels in 2010 to generate 1 gigawatt in peak sun. This is less than the new capacity of coal-fired power stations being installed in China every four days. All the world's manufacturers of solar panels added together are likely to be producing about 12 gigawatts of new panels a year by 2010, barely enough for six weeks of grid expansion in China.

So can PV ever become a technology that supplants a significant amount of fossil fuel generation? Demand for electricity is increasing by 3.5 per cent a year outside the industrialised countries. The small number of companies with the technology and experience to make competitively priced PV are going to struggle to make enough panels to cover the worldwide growth in electricity demand over the next few years. But we shouldn't be too pessimistic: the scope for wholly unexpected and truly revolutionary advances in photovoltaic technology is at least as great as any of the other technologies discussed in this book. If Nanosolar or one of its competitors does find a way of printing huge volumes of cheap semiconductor materials that can be easily added to the exterior of most buildings, the scope for photovoltaic technologies to change the world is almost unlimited.

Once the technical problems have been solved, the way is clear for PV. The environmental consequences of photovoltaics are limited and objections to the appearance of panels on the roofs of buildings or in large farms are few. The cadmium telluride used in First Solar's and some other manufacturers' panels is toxic but presents few dangers when in use in solar installations. Claims that thick silicon panels embed more energy than they are ever likely to capture from the sun occasionally resurface, but are not supported by the research carried out into the energy balance of PV. Studies some years ago suggested that panels repaid the energy cost of making them within about three and a half years, but advances in manufacturing efficiency and in the amount of light captured by PV mean that the energy payback period is now probably only about two years. Since the panels will normally last over twenty-five years, the return is good. Thin-film panels have even better energy balance because they require much less energy to make.

Concentrated solar power

Photovoltaic cells directly convert photons from sunlight into electricity. The other way of generating power from the sun is to concentrate the rays on to a liquid. The liquid heats up, and can be used to boil water, which then forces its way through a steam turbine, generating electricity. The efficiency of this process, expressed as the percentage of the sun's energy converted into electricity, can be greater than with a PV panel. Steam turbines are the method of generating electricity used in all existing coal and nuclear plants, so we understand well how to convert heat into movement and then into electrical energy.

This form of electricity generation is now usually called concentrating solar thermal power, often shortened to CSP. This name covers perhaps five or more separate approaches. The newest to arrive in a commercial application is known as 'solar towers'. At the first working example, near Seville in Spain, 600 mirrors placed in a circle reflect concentrated sunlight on to a single point at the top of a specially constructed tower. At present, this tower generates only about 11 megawatts – equivalent to two or three large wind turbines working flat out – but as more mirrors are added, the power will increase.

Another solar thermal technology is the solar dish. Looking like a huge satellite receiver from the 1960s, this apparatus tracks the sun as it goes across the sky. The mirrors on the interior of the dish reflect sunlight towards a focal point. At this point there is a Stirling engine, a machine for turning the expansion and contraction of gases

The solar tower near Seville

into power by turning a crankshaft. At the moment, the working examples are few in number and investors are still to be convinced of the financial potential of this approach to solar energy.

The best-established solar thermal technology is slightly different. This form of CSP uses long parabolic troughs covered with reflective material to concentrate the sun's powers on to a thin tube, called a receiver, in the centre of the parabola. A good solar collector can focus about a hundred times the normal power of the sun on to this. The receiver contains water or, more usually, oil. At the new large project near Granada, Spain, the thin tube of oil is heated to over 400°C in full sun. The hot oil is passed through water, with which it exchanges heat. The water rapidly heats up, boils, and then turns into energetic steam, ready for powering a rotating turbine, in exactly the same way as it would in a coal-fired power station.

The Granada power plant, called Andasol 1, is one of the first of what its German proponents hope will be tens of thousands of similar installations across the sunniest parts of the world. It is size-able, covering an area of about half a square kilometre. Several features help maximise its usefulness as a generator of electric power. The parabolic reflectors run in north–south lines and, using small electric motors, the mirrors are rotated from east to west during the solar day. This ensures that they will be facing directly into the sun during the daytime hours. The second major feature is the use of molten salts to store heat from the plant in order to extend the operating day to 17 hours or more. During the sunniest part of the day, part of the heat that is collected is used to melt the simple salts (potassium and sodium nitrates). The retained heat is then used to create the steam needed to power the turbines during the periods when the sun is down.

The new Andasol plant generates about 50 megawatts when it is working. Over the course of the year, it will deliver about 180 gigawatt-hours of electricity, providing enough for about 50,000 European homes. To put this in a slightly pessimistic perspective, we would need about 30 of these plants to provide as much electricity as we get from just one coal-fired power station. Though interest in this technology is growing rapidly, it is from a low base: one assessment suggested that only 30 large CSP projects were in active development around the world in early 2008. In September 2009, 800 MW of capacity was under construction in Spain and about ten times as much was in the early stages of permitting.

Many countries are extremely receptive to the Andasol-style approach to CSP. A similar power station was completed in February 2008 in the US state of Nevada. It also uses troughs of parabolic mirrors arranged in long rows. The Spanish construction company that built Nevada Solar One proudly says that it has 48 linear miles of parabolic collectors with 182,000 separate curved mirrors that focus the sun's energy on to 18,000 absorbing tubes. It covers an even bigger area than the Granada power station.

Very sensibly, Nevada Solar One was built very near to a long-distance power transmission line, meaning that it was relatively simple to connect to the electricity grid. In Nevada, and much of the southern US, the need for electricity peaks in summer afternoons, as air-conditioning is running at full power. The high level of demand generally means that spot prices for power are also at their maximum at this time. Electricity traded between companies in the wholesale market can be several times more expensive at this time than in the cool of early morning. Usefully, a summer afternoon

The Andasol concentrated solar-power plant

is when a solar power plant is also producing most electricity, meaning that its output commands a premium price.

Utility companies in the parts of the world facing power shortages on hot afternoons are likely to be particularly keen on CSP plants.

In February 2008, an Arizona electricity supplier announced a plan to build the world's largest single solar thermal installation, about 70 miles south-west of Phoenix, and a smaller Tucson plant was announced in early autumn 2009. The 280-megawatt Phoenix plant will be operational in 2011 and will triple the amount of renewable electricity produced by its owner. When fully operational, the plant will provide the electricity for 70,000 power-hungry Arizona homes.

By comparison with most other renewable technologies, CSP plants will not be particularly energy intensive to construct. But even the relatively simple solar collectors use a lot of steel. The Arizona plant – called 'Solana', Spanish for 'sunny place' – will use 80,000 tonnes of the metal. Made in an efficient blast furnace, the manufacture of this steel will cause the emission of perhaps 160,000 tonnes of carbon dioxide. It will take at least six months for the operation of the plant to pay back this carbon debt. Nevertheless, this is a better ratio than conventional PV panels and about the same as wind turbines.

Solar concentrators also have the enormous advantage of being relatively simple and reliable. The first CSP plants were built in California about twenty years ago and have worked well since then. The total output of these remarkable power stations in the Mojave Desert is six or seven times higher than that of Nevada Solar One or Andasol. They have a good record of reliability and are expected to last at least another fifteen years. Surprisingly, it is only recently that investors have come to see the advantages of replicating this successful experiment elsewhere in the world.

Perhaps equally importantly, CSP doesn't need scarce metals, so its growth won't be held up in the way that First Solar's photovoltaic cells made of cadmium telluride cells may be. It doesn't require expensive silicon, so it will escape some of the problems of conventional polycrystalline panels. It is almost completely non-polluting and, at least in theory, we can build multiple plants very quickly.

What is holding it up? Why are we not seeing hundreds of CSP plants in construction across all our hot deserts? Two reasons predominate. First, the current generation of parabolic dish reflectors is expensive. The total cost of the new Granada plant is about $8

Concentrated solar power in the Mojave Desert

per watt of peak capacity, while the Nevada station is about $5, both well above 'grid parity'. The solar PV manufacturers are aiming for a figure of about $1 for every watt of midday generating capacity and about another dollar for the associated electronics. CSP plants deliver much more energy in the morning and afternoons (because they can follow the sun, whereas most PV is fixed on roofs) but this comparison shows that CSP still has some way to go to be competitive with fossil fuel technologies.

Nevertheless, I think we can be optimistic that CSP will reduce in price at a similar rate to the falling price of solar PV panels. The troughs of parabolic mirrors are the largest portion of the cost of a CSP installation. Very large-scale manufacture of these parabolas will reduce their cost dramatically. As the world accumulates more knowledge of how to manufacture the other components of solar thermal installations, we can expect rapid decreases in their price. The mirrors have to be very precisely engineered to focus the sun's energy accurately on a thin tube of oil, but this is a manufacturing rather than a technological problem. My guess is that costs will respond rapidly as the CSP industry grows.

Some of the photovoltaic manufacturers are highly sceptical about the competing technology. They point to the need to install new electricity transmission cables to take electricity from remote deserts and also to the requirement to use large amounts of water for cooling. Water is not generally easily available in the hottest areas. Despite challenges such as this, the US Department of Energy gave an upbeat assessment of CSP in 2007:

> Existing CSP plants produce power now for as low as 12¢ per kilowatt-hour (including both capital and operating costs), with costs dropping to as low as 5¢/kWh within 10 years as technology refinements and economies of scale are implemented. Independent assessments by the World Bank, A D Little, the Electric Power Research Institute, and others have confirmed these cost projections. While not currently the lowest cost electricity, CSP is already close to competitive in peaking markets.

By 'peaking markets', the Department means the time when electricity is most expensive and most in demand: that is, late afternoons in summer in most of the southern United States.

A small group of determined scientists and engineers has been working to excite policymakers with a grand plan for CSP. Trans Mediterranean Renewable Energy Cooperation (usually known as TREC) is pushing for almost unlimited adoption of solar thermal power. Its backers include governments from North Africa and the Middle East. Prince Hassan bin Talal of Jordan, a country largely consisting of hot desert, was quoted voicing his support for it in the first few paragraphs of this chapter. TREC's view – and nobody has ever stepped forward to contradict it – is that putting huge developments of solar troughs in the sunniest parts of North Africa and the Sahara could provide all the electricity that Europe and the Middle East needs. The cost per kilowatt-hour, including a profit margin, would be competitive with electricity made from coal. TREC

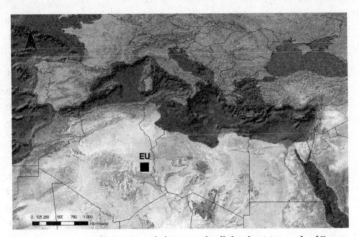

*The area needed to provide all the elecricity needs of Europe
from concentrated solar power*

mentions figures eventually as low as 4–5 euro cents, very similar
to the estimates of future costs from the US government. At these
levels, CSP plants are likely to be financially very successful. If we
had started pushing CSP twenty years ago, it would now probably
be producing electricity at a lower price than any other technology.
This should be a lesson for us.

TREC claims that the whole of Europe, North Africa and the
Middle East could get its electricity from CSP plants on as little
as 0.3 per cent of the desert land area of the region. In one of the
more telling illustrations of the power of the idea, TREC's website
has a map of the Sahara Desert on which is superimposed a tiny
red square. All the electricity for the entire European region could
come from power plants that would fit into this area.

But here we come up against the second problem faced by CSP.
Unlike solar photovoltaics, whose effectiveness degrades at high
temperatures, CSP will produce more electricity in the hottest
weather. But getting the power out of a hot and uninhabited North
African desert is likely to be difficult. A successful CSP plant needs
to be close to a high-voltage electricity line. The very sunniest areas
of the world, such as the Sahara Desert or some of the American
deserts of the south-west, have no high-voltage power cables at all.

The Sahara is a good place to generate the electricity, but we need a way of getting the electrons all the way to where they are needed – the populated areas of Europe, across the Mediterranean sea.

Moving electricity long distances is troublesome. Building the power lines is expensive, but a more significant problem is that a substantial amount of the electricity is lost in transmission. With conventional alternating current (AC) pylon lines, too much of the power would disappear as dissipated heat or as waste electromagnetic radiation on the way to the place using the electricity.

CSP enthusiasts have an immediate response to this difficulty. They say that we should be using high-voltage systems with direct current (HVDC) transmission. The losses in HVDC systems are much smaller than in conventional high-voltage AC systems. A transmission line from North Africa to England would lose less than 10 per cent of the electric power.

But are such transmission lines feasible? The electricity would have to go from, say, the Tunisian desert across the Mediterranean to Sicily and then northwards through Italy. The distances are long and the terrain will sometimes be extremely inhospitable. Getting to the Tunisian coastline should present few problems. Underwater cables would then be needed to cross the sea. The longest undersea power line today is the just-completed 580-kilometre link between Norway and the Netherlands. A huge cable weighing 80 kilograms per metre carries HVDC between the two countries, enabling their electricity grids to exchange power. The cost of this cable, finished in April 2008 and fully operational a month later, was about €600m, or about €1m a kilometre. The line allows 700 megawatts to flow either way, but the expectation is that most of the power will come from Norway's hydroelectric stations into the European power grid. A cable from Tunisia to southern Italy would cover a much shorter length, and the sea conditions would make installation much easier.

So the distances are feasible, though the cost of the cable running along the floor of the Mediterranean will be high. Nevertheless, the world's electricity industry is used to this scale of investment. In fact, direct current transmission lines above ground may be cheaper and simpler than the alternating current pylons running across Europe's landscapes today. HVDC pylons can actually be smaller and less visually obtrusive than their conventional AC equivalents.

New onshore HVDC links are getting longer all the time. India and China's need for electric power is surging and both countries are installing several major long-distance lines. In February 2008, the German power company Siemens completed a 1,200-kilometre HVDC link from power plants in western China to the industrial province of Guangdong. The pylons carry about 3,000 megawatts – more than the output of one of the very largest coal-fired power stations. Siemens is also building a long HVDC line carrying 5,000 megawatts, and its competitor, the Swedish/Swiss company ABB, aims to complete a 2,000-kilometre 6,400-megawatt link, also in China, in 2011.

How many pylon lines of this type would be needed to satisfy Europe's electricity demand with electricity from the Sahara desert? Germany's peak demand is about 100 gigawatts. Let's assume CSP from the Sahara eventually provides 50 per cent of the country's electricity demand, with the rest generated by local wind, tidal, and other renewable supplies. Germany would therefore need about eight of these long-distance HVDC links from the desert, each with a typical length of perhaps 3,000 kilometres.

To provide the whole of Europe's electricity would probably mean at least thirty different transmission lines of the same size as the biggest direct current links being built today. One estimate has them costing over €2bn each, including the portion under the Mediterranean. But, in itself, this is not an insuperable problem. A new nuclear plant may cost €5bn or even more. Nevertheless, analysing how the HVDC links could be built and financed is a challenge that has so far received too little attention.

The TREC concept is immensely attractive, and not just for Europe. Desert-based CSP could provide cheap, carbon-free electricity for the countries of North Africa and the Middle East. These states could use the power for industrial development and, perhaps most importantly, for the desalinisation of water, thereby allowing a major expansion in the area of irrigated crops as well as improved availability of drinking water. It may also be worth noting that the proponents of CSP believe that the areas underneath the parabolic troughs will be very productive places for horticulture. The shading effect will improve yields and reduce the otherwise excessively high temperatures.

Algeria is one of the first North African countries to back a CSP project. The Spanish construction company Abengoa, perhaps the most enthusiastic proponent of concentrated solar power, is building a power station there using parabolic troughs. This plant will be able to produce electricity for the entire day because the turbines can be switched to burn natural gas at night. This is a useful advance because it means that the solar plant can operate as what is called 'baseload', reliably producing a steady stream of power at all times and in all weather conditions. Most renewable technologies do not offer this security to the electricity companies. They are unreliable, such as wind power, or their power is cyclical. Tidal barrages, for example, generate most electricity just after high tides and none at all at low tide. So the Algerian plant is providing a useful prototype of how we can make CSP an integral part of the power grid. Algeria alone is talking about installing 6 gigawatts of CSP capacity, equivalent to three very large coal-fired power stations.

Most assessments of CSP agree that the costs per kilowatt-hour are likely to decline to below the figures for fossil fuel plants. A substantial carbon tax, applied to all fossil fuel power stations, is likely to improve the position still further over the next few years. But will concentrated solar power beat nuclear electricity on price? Optimistic forecasts see nuclear plants delivering power at 2 or 3 pence per kilowatt-hour. But these figures assume that the construction of nuclear plants can be done to the cost and timetable set out by the contractors. The experience at the new Finnish plant in Olkiluoto, discussed elsewhere in this book, gives us little reason for optimism.

The only real obstacle to generating most of Europe's electricity using North African solar collectors is the intimidating scale of the TREC project. To satisfy half the UK's electricity demand (or about 180,000 gigawatt-hours a year), we would need 1,000 plants of the size of the new Andasol facility near Granada, or 200 plants equivalent to the planned power station in Arizona. This is perfectly feasible – there are no obvious bottlenecks involved in the world's manufacturing industry scaling up over a decade or more to produce the reflective troughs that we need – but making it happen rapidly will require unprecedented international cooperation. In July 2009, a consortium of major banks, utility companies and technology businesses took the first step, linking up to form an international consortium to commercial North African and Middle Eastern solar power initiatives.

CSP could provide power for most of the world, not just Europe. The TREC project says that 90 per cent of the world's population lives within 2,700 kilometres of a hot desert. China could get a lot of its power from the Gobi desert, while the south-west US could conceivably produce enough electricity for the whole of the country. Australia, with its small population and enormous resources of sun, would be able to export energy to Asia. The distances would probably be too great for pylon lines but Australia could instead use its excess power to crack water into its constituents, hydrogen and oxygen. The hydrogen, a valuable fuel, could be exported by sea in pressurised tankers.

We cannot yet know whether CSP will turn out to provide cheaper power than solar photovoltaics. But CSP has the very substantial advantage of being based on simple and easily reproducible technology. PV technology is still the exclusive preserve of a small number of very secretive companies, all understandably eager to protect their intellectual property. This doesn't improve the chances that PV will grow fast enough to decarbonise the world's electricity production any time soon. CSP has more of a following wind: many companies around the world should be able to install relatively efficient power plants. The Spanish construction companies currently leading the world have fewer technological advantages over potential competitors than First Solar or Nanosolar have in the field of photovoltaics. In the language of venture capitalists, this means that the barriers to entry for new competitors in the CSP business are relatively low and we can expect substantial competition between firms striving to drive down costs. This is not to dismiss what Abengoa and the other Spanish companies already have done, which is technically very impressive. But CSP is more 'scalable' than solar photovoltaics and manufacturing capacity can increase fast as governments and companies get excited about the technology. Progress will be slower and more erratic than we might like but large-scale deployment of CSP will be able to provide much of the world's energy consumption within a few decades. We should try not to put all the CSP plants in a small number of countries, which might allow them to hold their faraway customers to ransom, but otherwise there need be no restriction on where the power stations are built.

Importantly, international grids of CSP power, such as the one proposed for Europe, need to be linked with electricity generating resources that can operate when the sun is not shining. North Africa has an average of seven hours per day of sunshine in winter and fewer than ten days on which measurable rainfall occurs. Cloudy winter days in North Africa are also likely to be the days when the wind is blowing hardest in the rest of Europe, but that relationship can't be replied upon. We can build North African plants that have heat storage, probably in the form of molten salts, but the international CSP grid will also need to be extensively linked to other sources of power, such as Scandinavian hydroelectric plants. These generating stations will have to be paid to be available at a moment's notice, ready to cascade water through their turbines if the sun ceases shining in North Africa. Other ways of handling short-term dips in electricity supply are discussed in Chapter 1. In the worst case, we will be able to use techniques for capturing carbon dioxide and link the CSP plants to natural gas pipelines to burn fossil fuels when the sun doesn't shine.

With complementary systems such as these in place and more enlightened thinking about how to match electricity supply and demand, concentrating solar power stations could provide a huge proportion of the world's power. It's a massively exciting technology that deserves much more attention from policymakers around the globe. With inexpensive photovoltaic panels also on the horizon, the world of solar-tech provides plenty of scope for climate and energy optimism.

Electricity from the oceans

Tapping tides, waves and currents

Go to the north-eastern tip of mainland Scotland and you reach the windswept and glaciated landscape of Caithness. As the land meets the sea, the nearby Orkney Islands can sometimes be seen across the eight-mile channel, often occluded by mist. Dangerous swirling currents and dramatic whirlpools make the waters a hostile place for all but the most experienced seafarers. White crested waves mark the places with the fastest currents. The guide for ships sailing in the area says that the water in these currents can be 'extraordinarily violent and confused'. The name of this narrow sea channel between the mainland and the Orkneys is the Pentland Firth, and it has one of the fastest tidal races in the world. Twice a day, immeasurable quantities of turbulent water shift back and forth between the Atlantic Ocean and the North Sea, containing a truly huge amount of untapped energy.

The Pentland Firth is one of perhaps twenty sites around the world – from the US and Canada to Australia and Indonesia – that promise enormous potential in terms of a relatively new technology called tidal-stream power. The idea is to position turbines on the bottom of the ocean to harness the enormous electricity-generating potential of these fast-flowing tidal currents. In most of the designs, these devices are, in effect, underwater wind turbines.

The wooden and stone windmills that dotted the hillsides of medieval Europe and Arabia were usually employed to grind wheat or other cereal grains. Today's wind turbines mimic the medieval windmill. Less well known is that our medieval ancestors also built mills driven by the ebb and flow of the tides. In a technical handbook on tidal power, written in 1921, a British Army officer, Major Struben, wrote that 'examples of such mills exist[ed] in England, on the Breton coast of France, in America and Spain, but, as far as can be ascertained, they were only of insignificant magnitude and primitive design, and, in consequence of their intermittency, not suited to ordinary industrial uses'.

This dismissive attitude to the usefulness of tidal energy was widely shared until recently. The total amount of energy in the tides across the world is not enormous, at least when compared to solar or wind power, or indeed the energy in waves. However, it is still far more than the total power needed by today's electricity grids. The energy contained in the global tides at any one moment is probably about 3,800 gigawatts, or two or three times today's worldwide electricity consumption. Most tidal energy is impossible to extract; it is found in deep oceans far from coastlines. But at a small number of places, such as the Pentland Firth, huge resources of energy are concentrated into narrow funnels.

Of course, tidal-stream turbines are not the only way to capture energy from the seas. Barrages are another option. These large dams harness their energy from the 'range' of the tide, or the difference between its high and its low points. The barrage is built across a tidal river or estuary, and the incoming tide is allowed in through sluices. When the tide reverses, the sluices are opened and the force of outgoing tide turns electricity turbines. We know that tidal barrages will work as there are already commercial plants in France, Canada and Russia.

There are also at least three marine-energy technologies that don't rely on either the range or the current of the tides. First, turbines like the ones already described could be positioned to collect the energy of the main ocean currents, such as the Gulf Stream. Second, wave power collectors can utilise the up and down motion of the sea as the waves pass. Finally, heat pumps can use differences between the temperature of the sea surface and the deep ocean to drive an engine, usually to generate electricity.

The Pentland Firth seen from the Caithness coast

All of these technologies are commercially interesting, but, as this chapter shows, the power from tidal currents and ocean waves looks the easiest to exploit and offers us the biggest potential for generating electricity.

The potential for energy from the seas

Although both the vigour and the regularity of marine energy have been obvious since mankind started to sail the oceans, we have been slow to exploit their potential. Even now, only a dozen or so sites around the globe successfully generate electricity from the oceans. France built a large barrage across the river Rance to collect energy from the tides on the northern coast of Brittany over forty years ago, and a small number of other places with large ranges between the high and low tides have installed similar dams. A prototype power station in Hawaii has occasionally generated electricity from ocean temperature variations. But the general picture is of hesitant and slow progress. Only in the last few years has the pace of installation started to pick up. The first two commercial-scale turbines that capture the flow of the tide on the ocean floor have been connected to the UK electricity grid and a small wave power farm has been installed off the coast of Portugal. Clean Current's prototype tidal-stream turbine has been successfully trialled at Race Rocks off the coast of British Columbia and a few other developers have put working machines in the water.

Why has progress been so slow? Until the recent upsurge in prices, cheap fossil fuels reduced the incentive to spend the large sums of money necessary to develop and construct devices that could profitably capture marine energy. At times of uncertainty about energy supplies in the 1970s, governments funnelled some money into marine energy research programmes but the interest faded as oil started flowing freely again. Several other interesting new technologies, such as solar photovoltaic power, went through the same cycle.

None of the government-sponsored engineering trials into marine energy thirty years ago provided a definite promise of competitively priced electricity. This was particularly true when oil was only $20 a barrel. Understandably, private capital has been slow to flow into untried technologies that looked as though they would produce electricity only at twice the cost of fossil fuels. Many intriguing but undeveloped ideas have been abandoned over the years as the money dried up.

One 2006 report suggested that the UK – with far more than its fair share of the global potential – had spent about £25m in total on marine energy research and development since 1999, some of which went on surveys and policy statements rather than engineering. This is little more than £4m per year, or less than a fifth of the annual subsidy received by the Royal Opera House in Covent Garden, London. Similarly tiny sums have been spent in Canada and elsewhere.

This hesitancy is slowly beginning to change. As oil shortages, energy insecurity and climate change raise everyone's fears, private venture capital and government R&D funds have started to arrive in larger amounts. Most of the UK's large electricity generators have begun to invest in marine energy, and have tentatively linked their names to the entrepreneurial small companies risking everything in the seas around Britain. These big businesses are getting involved in wave and tidal power partly because they need to show an active interest in renewable energies; all socially responsible energy companies recognise the importance of demonstrating, as visibly as possible, their commitment to renewable electricity generation. However, this is not the only reason. Large energy companies have long memories, and they can see that the small number of firms which bravely committed to wind power fifteen years ago are now reaping the returns on their risky investments. The UK has more usable wind power than anywhere else in Europe, but neither the government nor the energy companies successfully nurtured a domestic industry. Today none of the world's big turbine manufacturers has a substantial presence in the country. Britain's coastlines provide a similarly powerful source of energy and this time the large electricity companies are less willing to see the opportunity escape.

So far, the sums invested by commercial investors are quite small. Few large companies quoted on the stock market want to put tens of millions of pounds into risky new ventures that may never produce a commercial product. There have been some cautionary recent examples of embarrassing failures that help to make investors nervous. One unlucky small company saw one of the blades break on its underwater turbine in the Hudson River near New York, while another entrepreneurial company saw one of its devices sink to the ocean floor off the Canadian coastline.

Nevertheless it is increasingly clear to investors that the UK and a number of other countries, such as South Africa, Australia, Portugal and Chile, have impressive resources of energy around their coasts.

The British Isles, including Scotland, possess at least 10 per cent of the world's accessible tidal energy, much of it in several extremely rapid races such as the Pentland Firth or around the island of Alderney off the coast of France. Britain also has the potential for several large barrages across estuaries and coastal bays. The tidal range of the estuary of the River Severn is one of the largest in the world and, more importantly, the water can be relatively easily trapped behind a barrage and gradually released to drive turbines. A third advantage is that the average height of waves around the UK coastline is high in comparison to most other coasts and so power is easier to extract. Entrepreneurial activity is not restricted to the UK. One example is Australia's BioPower, a business making both tidal current and wave energy collectors. Its unusual-looking devices (pictured) mimic the motion of fish moving through water or the waving of weeds on the sea floor.

Even though marine energy looks a good prospect, no one should pretend that funnelling government or private money into marine energy in the UK or elsewhere is going to be a guaranteed success. Of the forty or so separate devices for collecting tidal energy currently in development in Britain's engineering laboratories and universities, how many will survive the first few weeks of commercial trial in fast-flowing seas? Perhaps a handful. Some failed projects will be able to transfer useful knowledge to other companies but most of the investment in marine technology is going to be wasted in fruitless endeavours.

Some of the forty designs look implausible even to a non-specialist eye. But we are at the stage in the development of the marine renewables industry when large numbers of competitors jostle for success and we have no means of telling which are going to work. These companies need support from government if they are ever going to be able to pay the multimillion-pound costs of designing, building and testing sea-going machines. Of course, government-funded R&D has an almost absurdly poor financial record in most countries, and this has made governments understandably shy of handing cash over to starry-eyed engineers with poor financial skills and personality problems. Nevertheless, it looks as though substantial financial help will be needed for early-stage research,

BioPower's wave-energy collectors (above) mimic the swaying of plants on the seabed. The company's tidal current generator (below) resembles the tail of a fish.

with private capital reserved for technologies that look closer to commercial launch.

Despite all the challenges, it seems entirely plausible to suggest that a small number of successful marine-energy companies will each be able to install thousands of robust turbines. In the UK, these should be able to provide 20 or 30 per cent of the UK's electricity within twenty years. Similar figures are possible for other countries with long coastlines. It took this length of time to move from a few lonely windmills erected on land owned by idealistic eccentrics to today's position, in which large generating companies are aggressively bidding to install major wind farms across the world. Costs have fallen several-fold in this fifteen-year period. Marine power has a similar potential to provide a substantial fraction of the electricity supply of coastal countries at prices no higher than coal or gas.

Tidal-stream energy

Tides are caused by the gravitational pull of the moon and, to a lesser extent, the sun. As the earth rotates, water is pulled towards the orbiting moon, resulting in small bulges in the water level. Most places experience two tides a day, though at some points on the earth's surface there is only one and at other places three or even more. The lunar cycles cause the height of the tide to vary over a period of two weeks. Spring tides – which are nothing to do with the season of the same name – are much larger than neap tides. At times of spring tides, the moon is aligned with the sun, combining the gravitational effects of the two and creating a greater pull on the oceans. At times of neap tides, the sun is perpendicular to the moon and counteracts its effect, thereby minimising the range of the tide. Additionally, tides are larger at the spring and autumn equinoxes than during the rest of the year. So although tides are entirely predictable and reliable, the amount of energy available to extract varies substantially from week to week and by several orders of magnitude from place to place. The world's tidal energy is concentrated on a relatively small number of coasts and bays, but at these places huge amounts of power are there for us to capture.

We cannot try to control the energy of a place like the Pentland

Firth by constructing a dam across this violent and deep stretch of water, extensively used by vessels going from northern Europe to America. We need to find ways of installing devices on the sea floor that will use the power of the moving tide. Although the technical challenges are more demanding than those involved in putting a barrage across an estuary, the available energy yield from tidal races around the world is much greater.

The Pentland Firth has not been studied in sufficient detail for us to be sure of exactly how much energy it contains. The very ferocity of its currents makes it difficult to measure water speeds with any accuracy. But most studies of the area suggest that this thin channel can generate 8 gigawatts when the tide is running at its peak. Louise Smith, a civil engineer with wide experience in road and viaduct construction around the world, was recently tempted back to this remote corner of north-western Europe after twenty years away. Her new job is to encourage the commercial exploitation of this enormous resource of energy. She is highly optimistic about its potential, and told me that some research suggests that the power in the tides of the Firth could be as much as 20 gigawatts – enough to comfortably cover the whole of London's electricity needs.

Until recently, the Pentland Firth was simply too intimidating a location for businesses even to contemplate developing machines to collect its energy. But just as the oil industry has moved into ever more inhospitable terrain as energy prices have risen, so are pioneers beginning to rise to the challenge of this most formidable of environments, hoping to harness its dangerous but enormous energy potential. In the next few years, several companies are hoping to collect energy from the tides in the Firth by planting devices on the seabed itself, either tidal turbines used singly, or ten or twenty such machines spread out in impressive array.

The power in the tides at such places is very dense, particularly when compared with wind power. Although the speed of the wind can be several times the maximum velocity of the tide, water is about a thousand times as heavy as air. As a result, the power available is many times greater. The energy in the flow of the Pentland Firth can be as much as 175 kilowatts in a vertical square metre. For comparison, the typical electricity use of a house in Europe is about half a kilowatt. So 350 houses could be powered by the peak energy of 1 square metre of tide ripping between the open Atlantic and the North Sea. Properly located, a relatively small tidal collector could, at least in theory, produce more than the largest windmill. In a powerful tidal race hundreds of turbines would function much as a wind farm does, collecting a good fraction of the total energy of the current.

Those backing the technology claim that the environmental impacts of a tidal farm will be extremely limited. The blades rotate slowly and are unlikely to pose much of a hazard to marine animals. The water is not trapped behind the dam (as is the case with a tidal barrage), so the ecology of the area will not be significantly affected. Perhaps these assurances are too glib – we have yet to see the effects of a full-size tidal turbine farm – but the scale of any environmental damage is probably going to be small. The best sites for tidal turbines are usually inhospitable places for fish and other creatures. The proposed Canadian development in the Minas Basin is characterised by a seabed scoured clean by the rapid flows of water, with some dunes of underwater sand and gravel.

Three or four of the world's most exploitable fast-running tides lie around Britain's coast, which looks like good news for the country. But it will be extremely difficult to solve the engineering problems involved in installing tidal-current generators. The force of the tides in the best locations is so enormous that machinery has to be built to extremely high specifications, while the salt and other minerals in the sea water will degrade and corrode all but the most resilient structures. The UK is lucky in that it already possesses one of the world's best-established offshore oil industries, one that is well practised in providing the highest possible mechanical reliability in the face of violent seas. Since the advent of deep-sea oil production thirty years ago, we have seen improvements in designing and fabricating devices that can last for decades in unforgiving environments, and major technological advances have reduced the risk of rust, weed infestation and water ingress.

Lunar Energy, a company that has shown some of the most exciting signs of technical and commercial progress, uses designs that come straight out of the oil business. Built in Aberdeen, the centre of

The Lunar Energy turbine depicted on the sea floor

the UK's offshore industry, Lunar's huge yellow turbine is designed to sit on the sea floor. To remain in the right place when the tides are flowing strongly, the device has to be heavy. The 1 megawatt version weighs 2,500 tonnes, equivalent to 60 fully laden lorries. Most of the weight is provided by inexpensive ballast, present simply to hold the turbine in place. The machine is 25 metres long and 15 metres high. Most of the UK's most powerful tidal currents occur in seas deeper than 40 metres, so the Lunar Energy turbine will sit well below the surface of the sea, minimising any danger to shipping.

The unusual shape of the device – a long tube that narrows in the middle – helps focus the power of the tide. Water flows into the tube and then is forced to speed up as the aperture narrows. Once past the turbine, the tube opens up again and the water slows to the same speed as the external current. In this way, the force applied to the rotors is even greater than for a turbine that simply uses rotating blades. The rotor rotates at a sedate 20 revolutions per minute, helping to minimise the wear on the moving parts. One of the many innovative features of the Lunar designs is that the rotating blades do not themselves generate electricity. Their movement forces hydraulic oil through a turbine above the main chamber. All the critical components in this impressive machine can be easily removed by a boat moored above the turbine. Lunar claims that this can be conveniently done in the quiet time between tides.

One of the first tidal farms in the UK will be put on the seabed off the Pembrokeshire coast, on the south-western tip of Wales. In cooperation with the huge German power generator E.ON, Lunar will install eight turbines by 2011, enough to power 5,000 homes. In March 2008, Lunar announced its first export order, for a planned 300-turbine tidal farm off the South Korean coast, to be completed by 2015. The machines will be built in Korea by a shipbuilding firm, avoiding the need to move these massive structures around the world. Lunar Energy gives very optimistic forecasts for the cost of electricity produced in areas of strong current, promising that once it has driven manufacturing costs down, this could be as low as 2.5–3.5 pence per kilowatt-hour, far less than carbon-intensive coal.

A large number of other firms from the British Isles are contenders for commercial success but two businesses have attracted particular notice. MCT builds twin-headed windmill-like turbines. The structure is supported on a single steel pile that has been driven into the seabed. Unlike the Lunar device, MCT's machine can be raised above the sea surface for maintenance. The disadvantage of this is that a portion of the structure is always above the sea surface, meaning that it is potentially more of a risk to marine traffic. The first large-scale installation of the MCT turbines was placed in Strangford Lough in Northern Ireland in March 2007, six months late because of an agonising wait for access to one of the small number of specialist ships designed for deep-water installation projects. (The UK's potentially huge offshore wind industry will probably also be held up by this crippling worldwide shortage of the vessels that can carry out work on installations of this sort.) A plan for a seven-turbine farm using MCT devices off the coast of North Wales is backed by RWE, the second-largest German utility. One of the crucial reasons why this location was chosen, apart from its tidal

The MCT turbine installed in Strangford Lough

speeds, was the availability of a nearby connection to the electricity transmission system. One of the problems with tidal energy around the world is going to be connecting the turbines to the power grid. The best tidal locations tend to be far from high-voltage transmission lines and to use the full force of the Pentland Firth, the UK will need a new offshore undersea cable running down the east coast to London and on to the rest of Europe, using the same HVDC technology that will bring Saharan solar power to Germany and other countries, as described in Chapter 2.

The third high-profile competitor for the wide-open market for tidal power generators is the Irish firm OpenHydro. This company produces a striking O-shaped device, consisting of a central rotor that spins inside an outer ring, generating electricity as it moves.

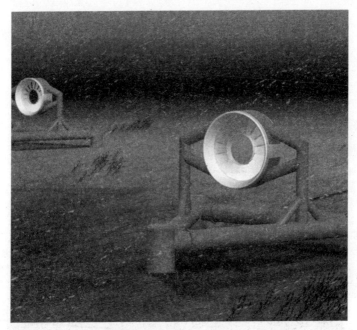

The innovative tidal-stream turbine from OpenHydro, with just one moving part

This extremely elegant system requires only one moving part and, as with the Lunar Energy turbine, the whole structure sits well below the surface, posing little risk to shipping. The first commercial OpenHydro devices will be installed in the tidal races off Alderney in the Channel Islands during 2009. This small island, lying near the coast of France, but with close ties to Britain, has tidal flows that rival the Pentland Firth for their concentrated power. One estimate suggests that peak flows at Alderney would be equivalent to 5 per cent or more of UK electricity use.

Another productive location will be the Bay of Fundy, an inlet between the Canadian provinces of New Brunswick and Nova Scotia. The Bay of Fundy, which is 180 miles long and 60 miles wide, has probably the largest tidal range in the world. On rare occasions, high spring tides can be almost 20 metres higher than low tide. The twice-daily water flows in and out of this channel are greater than all the rivers and streams in the world combined, making it an obvious target for power generation. One previous attempt to exploit a very small fraction of the power of the water was made in 1984 by constructing a tidal barrage across a river flowing into the bay. The interest in tidal power has increased in recent years and in late 2009, the government announced a plan that will see three different types of tidal turbine put into the waters of the Minas Basin at the eastern extremity of the Bay of Fundy. The goal is eventually to use the huge forces of the tide in the Bay to power all of the 800,000 homes in Nova Scotia as well as parts of New Brunswick and Prince Edward Island.

On the other side of Canada, the province of British Columbia has identified ninety separate sites that have enough tidal current to make it worthwhile to extract energy. Most are close to Vancouver Island, the main centre of electricity demand, meaning that exploitation here is particularly attractive. These sites could produce 4 gigawatts, a sizeable fraction of the total demand from the whole of British Columbia. As in the British Isles, the local availability of tidal energy has spawned the early beginnings of an industry trying to commercialise designs for tidal power. Clean Current, a Vancouver firm, has had an experimental turbine of the same style as OpenHydro in the waters of British Columbia and is involved in the bigger trials in the Bay of Fundy.

The tidal-stream generator designed by the Australian firm BioPower is based on copying what the company calls 'the highly efficient propulsion of Thunniform mode swimming species, such as shark, tuna, and mackerel'. The first commercial prototype of this device will be installed off Flinders Island, Tasmania, in 2010, supplementing the diesel generators and wind turbines that provide the small community's electricity. The company hopes to have a 250-kilowatt machine for sale by 2011.

Predicting the success or failure of technologies at an early stage in their development is not a game played by sensible people. In the case of tidal currents, forecasting is even more difficult than usual because of the multitude of different turbines in development. It is impossible to predict which will succeed. But will one or more companies manage to develop a turbine that reliably and inexpensively generates electricity at significantly less than 5 pence per kilowatt-hour, making it broadly competitive with fossil fuels? I think the answer is almost certainly yes. We do not have to challenge the laws of physics or those of thermodynamics as we have to do with some other technologies. The crucial problems are those of mechanical engineering and are thus more susceptible to eventual solution, probably by continuing to improve strength and robustness.

Barrages that utilise the range of the tides

To those of us with childhood memories of huge English Channel beaches gradually laid bare by falling tides, the average difference around the world between high and low water seems surprisingly small. The typical figure is somewhat less than a metre, compared with 5 metres or more around the British Isles. The largest tidal ranges in the world are found on the eastern coastline of Canada in the Bay of Fundy and northwards into Labrador.

In places where the difference between high-tide and low-tide levels is large, the most obvious way of capturing energy may be to build a barrage rather than installing tidal-stream machines on the sea floor. Barrages usually work by letting the incoming tide flow freely though an embankment. When high tide is reached, and a large body of water is sitting behind the dam, the gates that allowed the water to flow through are shut. From this point on, the barrage works in exactly the same way as a hydroelectric power plant. As the tide falls, reducing the water level outside the barrage, a height difference develops between the water levels behind and in front of the barrage. Water can then be allowed to flow through the dam towards the lower level, turning turbines as it goes and generating electricity. The best sites for barrages are likely to be enclosed bays or large estuaries.

Tidal barrages could, in theory, generate electricity when the water moves through the turbines in both directions, inwards as well as outwards. But in the small number of barrages currently operating around the world, it has not proved easy to capture energy at both phases of the tide. This means that the barrages operating today can only produce electricity for about half the day. The amount of power generated will peak a couple of hours after the high tide has passed and will then fall away.

Spring tides, which typically have double the range of neap tides, will provide far more power at tidal barrages than neap tides. So the amount of power available to a national electricity system will vary in a regular cycle. Very usefully, however, the peaks of spring tides always occur at the same time of day at any particular location. So, for example, we know that most power from the proposed barrage across the River Severn in England will come between 1 and 3 p.m. But other places around a coastline will have their peak at different times. This means that a portfolio of tidal barrages and turbine farms spread along a long coastline will potentially provide very stable levels of power throughout the day. Tidal power can thus avoid some, but not all, of the intermittency associated with wind and solar power.

We will be able to forecast reasonably accurately how much power will be delivered every minute, years in advance. That said, as with tidal current power, the amount generated will vary through the seasons, with the equinox tides being more powerful than at other times. Power generation will also be slightly affected by the weather, since strong winds affect the height of tides, though this effect is likely to be quite minor. All told, a wide spread of tidal generators will be almost as useful to the operators of an electricity grid as a coal-fired power station.

The river Rance dam has been successfully generating a peak 240 megawatts (enough to meet the electricity needs of almost half a million French homes) for several decades. The best-known scheme for a much larger tidal barrage is the one proposed for the Severn, the major river draining much of Wales and parts of western England. The Severn has a huge tidal range – probably second only to a few places on the east coast of Canada – because the tide is funnelled up a sharply narrowing estuary. A single dam across the river, about 16 kilometres in length and costing about £15bn, would generate roughly 5 per cent of the UK's electricity consumption. It is not all good news, however: a UK government body has estimated that if constructed by private capital the project would produce electricity at a cost of over 9 pence per kilowatt-hour, much higher than many other types of renewable technology. Other dams around the UK coastline could provide another few per cent of the country's electricity, but possibly at an even higher cost.

In addition, tidal barrages have some unfortunate side-effects. They change the ecology of the areas behind the dams because they reduce the range of the tide. The Severn barrage would also cause high levels of fish mortality, since many of the unfortunate creatures would be sliced into fillets as they passed through the turbines. In addition, the movement of silt is impeded, significantly affecting the ecology of large areas behind the dam and probably disrupting the colonies of birds that feed in the estuary. However, it is the high price of most tidal barrages that really upsets their viability. While the cost of, say, mechanical turbines falls as we construct more and more of them, the same is not likely to be true of building huge one-off concrete dams many kilometres long.

Another important reason for scepticism about the value of tidal dams is the relatively small number of locations where significant amounts of power are available. It might be well under a hundred across the globe. So although the UK might focus on the potential for the Severn estuary, elsewhere the scope for using tidal power is relatively limited, except for some promising sites in Russia and Canada. For all these reasons, tidal barrages remain a less exciting technology than tidal-stream turbines, which promise fewer side-effects and a greater potential for plentiful, inexpensive power.

Power from the waves

Tidal energy is generated by the tug of the moon and sun, but wave power comes from the lashing of the oceans by winds. Wave power is therefore a much-mutated form of solar energy. The sun heats the air and temperature differentials between areas produce wind, which then generates waves when it passes over sea. Less than 1 per cent of solar energy becomes wind, and a very small percentage of wind power is then transformed into waves. These facts would appear to make it less attractive to collect wave energy than solar power or wind. But in some ways waves are better. The crest of the waves in the open Atlantic seas off the coast of Portugal and the Pacific seas of Washington State might contain as much as 70 kilowatts in each square metre of sea. If captured, the power from this tiny area would be enough to heat ten large and draughty houses in the depths of a very cold winter or provide the typical electricity needs of well over a hundred homes. By contrast, a square metre of hot tropical desert might only receive a kilowatt of solar energy even when the sun is high in the sky.

Tidal power is concentrated in a small number of areas, but wave power is widely spread around the globe. And though the best tidal sites may contain more energy, wave power is available for hundreds of kilometres along straight shorelines. The best areas tend to be in temperate zones, where western coasts are exposed to the prevailing winds and where frequent strong storms whip up high waves. Very high levels of available energy have been measured off coastlines as diverse as southern Chile, western Australia, Portugal and South Africa.

Nevertheless, we will never find it worthwhile to collect more than a small fraction of all wave energy. The largest waves are found in open seas, hundreds of miles from coastlines. We might be able to install energy collectors there, but transporting it to the nearest electricity grid would be costly. Moreover, there is an important design dilemma that affects almost all wave collectors. The stormiest seas contain truly awesome amounts of power. If we tried to collect this energy, the devices would have to be extraordinarily strong, able to resist ferocious forces, occasionally exceeding what is seen even in the most powerful tidal currents. To withstand these forces, the machines would have to be so robust that the cost of manufacturing them would be crippling. Much like wind turbines, which close down in gales, most of the wave collectors currently being tried out around the world don't attempt to operate in storms. This means that they don't harvest energy during the times when the waves are at their most energetic. In this and other ways, wave collectors tend to be designed for survival and not for maximum energy output. The Australian company BioPower, whose fish-fin tidal-stream generator I mentioned above, has also developed a wave power device. Its prototype, which mimics the actions of water weeds in turbulent water, does not try to resist the force of the waves but simply lies flat on the ocean floor when the energy of the seas is too violent.

Even if they harness a few per cent of the passing energy, wave devices are capable of filling a large fraction of the total energy needs of many countries. For example, the British wave industry trade body claims that the total amount of accessible wave energy in UK waters is about twice the country's total electricity use. (The leading British wave power developer, Pelamis, gives a similar figure but qualifies it by saying that only a portion might be economically recoverable.) The industry trade association also quotes analysis that suggests the worldwide availability of electric power from the waves might be as much as four times current global electricity use. Other sources claim even higher figures. Whatever the correct number, wave power should be able to service a large fraction of our needs for electric power.

A host of different wave energy approaches are jostling for the attention of banks, governments and electricity companies. One industry website reports that a search of the patent literature throws up more than a thousand proposals for wave energy collectors. But, as with tidal energy, we can be sure that most of these collectors, if built, would be in pieces on the sea floor within a few hours of the start of a severe storm. Indeed, only a handful of prototype wave devices have ever exported power consistently. Nevertheless, increasing interest in renewable energy, higher levels of government funding, rising electricity prices, and better construction techniques have combined to make wave power an exciting area to watch.

The first commercial wave farm in the world lies 5 kilometres off the coast in northern Portugal. At this distance from the shore, waves are much more powerful than they are nearer the coast, since their energy dissipates as they run up to the beach. Swell from the Atlantic is captured by three Pelamis machines, made in Scotland but hooked up to the Portuguese grid. Generating about 750 kilowatts each when the wave conditions are right, these long red articulated cylinders are the fruit of over thirty years' work. The story begins in the engineering workshops of the University of Edinburgh. It was in 1974, just as the world was waking up to the potential volatility of energy prices after the Arab oil embargo, that Professor Stephen Salter came up with the idea of a device that productively absorbed energy from the waves. Even then, it was obvious that a wave collector could potentially capture 80 per cent or more of the power of a wave: this is far better than is ever likely to be achieved by a photovoltaic device converting the energy of the

The Pelamis wave collector during its sea trials

sun, and a higher percentage than the theoretical maximum that can be achieved by a wind turbine.

Photographs from the period show engineering students in the 1970s employing arrays of analogue electronics to control the waves in a tank and to measure the energy captured by what became known as Salter's Duck. The 180-metre-long red Pelamis machines that float semi-submerged in the powerful waves off north Portugal are the very indirect descendants of this work, benefiting from many generations of prototypes and thousands of hours of testing in the wave research centre in the Orkneys. These huge structures, weighing 750 tonnes each when filled with ballast, are the nearest thing the world has to proof that wave energy can be profitably extracted. They've had mechanical problems – and the majority owner of the wave farm fell victim to the 2008 financial crisis – but the technology has proved itself and German multinational utility E.ON has placed an order for the next generation of Pelamis devices to be installed off Scotland.

The Pelamis is composed of four cylindrical segments, each joined to the next by a flexible link in which the power generation module is found. The whole snake-like device is loosely moored on the sea floor and aligns itself automatically at 90 degrees to the prevailing wave direction. As the wave passes along the Pelamis, individual segments rise and fall. This motion causes the joint between the cylinders to flex, pushing hydraulic rams which pump oil under pressure. This pressure is converted into electrical energy via a turbine. The electricity from multiple machines is then combined and sent onshore. As with many of the other interesting wave capture devices, the machines are designed and built by people with experience of constructing massive steel structures for the Scottish offshore oil industry. They are engineered to last decades, but if they need maintenance these steel sea snakes can be unhooked and easily towed into port for repair.

Portugal has excellent wave power and its shores are an obvious choice for the first attempt to put multiple machines into the water. Atlantic westerlies whip up the seas along the coastline, generating high and relatively reliable waves. In September 2009, Energias de Portugal, one of Europe's major electricity utilities, bought out the existing owner of the first three Pelamis machines and all is now set for them to be followed by twenty or thirty others, spaced over a square kilometre of ocean. The company commercialising the Pelamis says that the technology could generate over 30 megawatts in this area. This is the same amount of power as six of the very biggest onshore wind turbines, which would probably occupy a much larger area of sea.

The price initially paid for the three Pelamis generators was about £6m, a substantial premium over the cost of a wind farm of the same power. But wave power is likely to be important to Portugal, which has no significant fossil fuel resources, and this investment is small in the context of the size of the opportunity. Wind generation is growing fast, but will eventually slow as the country runs out of good onshore locations. One prediction is that waves will produce almost a third of the country's electricity by 2050. The electricity company that constructed the Pelamis wave farm will want to capitalise on its early experience and capture itself a substantial fraction of this market.

The second crucial reason why Portugal is the first place to install a working wave farm is the price that the electricity companies are obliged to pay for power generated from the sea. Government regulations mean that the owners of the Pelamis machines will get about 18 pence (23 euro cents) per kilowatt-hour, several times the standard wholesale price for electricity. When the farm is working at its peak rate, it will be earning £400 an hour. Even at half capacity, the three machines will produce over £1.6m worth of electricity in a year, providing a reasonable return on the initial investment. But without the high prices available as 'feed-in tariffs', the Pelamis would be far too expensive. The first machines went to Portugal because the level of support from the UK government is much lower.

However, once the costs of the Pelamis fall, or the price paid for wave-generated electricity rises, the rush for wave power will start across the west coasts of the United Kingdom and many other places around the world. An area of 1,000 square kilometres could potentially provide over a third of the UK's total need for power. One suggestion is to create a long strip of wave farms 2 kilometres wide around the exposed western coasts of England, Wales and Scotland, taking the electricity onshore at points where the local grid connections are strong. Even though the scale of such a wave farm would be enormous, it is unlikely to change local ecology much. Although the Pelamis is very long, it sits only a couple of metres deep in the water and seems to offer minimal disruption to flows of fish and sea mammals. It seems logical to install banks of devices such as these intermingled with the huge offshore wind farms that are very likely to be built in similar places.

Other countries with substantial wave resources could be significant beneficiaries once wave collectors have reached commercial viability. Take South Africa, where electricity demand is growing fast, but production has been failing to keep up, with blackouts repeatedly shutting the gold mines and costing the country billions in lost exports. Wave farms at South Africa's southern tip could provide at least 20 per cent of the country's power needs, reducing the shortfall without raising carbon emissions. In fact, almost all countries outside the tropics with long west-facing open coastlines could generate a large fraction of their power from the deployment of wave collection devices.

Pelamis's sponsors think that their product has a good chance of eventually producing electricity at prices equivalent to or lower than those of fossil fuels. Independent estimates of future costs are falling too, as analysts see that large-scale manufacturing will improve prices. But there have been significant delays in commissioning the Pelamis machines in Portugal, and we cannot know yet whether this particular approach to collecting wave power will be the one that gets adopted around the world. Investors have already put £40m into the company developing the Pelamis, a huge sum by the standards of the European renewables industry, though trivial in the context of the money that has gone into solar photovoltaic or cellulosic ethanol plants in the US. Investors must be hoping that the long wait before their machines can be used is almost over.

What if the Pelamis turns out not to work, or to be too expensive to be widely used? What is the next most plausible type of wave power collector? Most attention is focused on devices that bob up and down in the water like fishermen's floats. This heaving motion drives a piston that moves a pressurised column of fluid inside the device, which in turn provides the power to drive a turbine. Small buoys in harbours often use similar technologies – but on a much smaller scale – to power the light that alerts ships to their presence. Canada's Finavera and New Jersey's Ocean Power Technologies are among the companies hoping to apply this technique for large-scale generation. Growing interest in harnessing the power of the waves off the west coast of the US has begun to help these businesses gain attention, but their technology is still at an early stage. In late 2007, one of Finavera's prototype buoys sank after a two-month ocean trial. Reports at the time suggested that seawater had got into part of the mechanism after a pump had failed. Despite this setback, Finavera went on to win a contract to build wave devices off the coast of Washington State and to get support from California's largest power generation company for another project.

A third type of wave generator also offers some potential. Many harbours and beaches have a breakwater: this is essentially a wall in the sea designed to minimise waves and create stiller waters.

Breakwaters can be designed to catch some of the water from the waves as they crash into the stones or concrete. Once trapped, the water escapes downwards through a shaft, and can drive a turbine, much like a hydroelectric power plant. The mechanical challenges of making such a device are minor since the breakwater itself has taken most of the force out of the waves, but the consequence is that relatively little energy remains to be used. These devices will be used in coastal protection schemes in the future, but are unlikely ever to generate a substantial fraction of world electricity need.

More generally, though, wave power offers enormous potential. The mechanical engineering challenges are probably as substantial as those for tidal current turbines but they're clearly solvable within a few years. Indeed, they're almost trivial compared to those faced by some parts of the offshore oil industry. All that's needed is support from governments – a high guaranteed electricity price and continued funding of university research – and help from large energy companies and private investors. With that support in place, a Pelamis-like device will eventually produce cheap electricity on a large scale around the world's western coasts.

The Gulf Stream and other ocean currents

The Gulf Stream isn't a tide, although it has some of the same useful features. It is a continuous flow of water moving from the Caribbean to the northern Atlantic, where the current sinks and returns south. This circular motion is part of the system of worldwide ocean currents, driven by the winds and by differences in water density around the globe.

The easiest place to exploit these currents lies off the coast of Florida, where the majority of the kinetic energy of the Gulf Stream is funnelled into a zone just 100 kilometres wide. Here, turbines that resemble the MCT tidal collector at Strangford Lough could be used to extract some of the power for conversion into electricity. At one point the Gulf Stream runs only about 25 kilometres off the Florida Coast, meaning that connection to the grid shouldn't be a major problem.

At about 8 kilometres per hour, the Gulf Stream moves much more slowly than the tidal races in, say, the Pentland Firth. This is significant, because the energy in a moving stream of water (as with wind) goes up exponentially with the speed – a stream moving at 16 kilometres per hour has eight times the force of a water of 8 kilometres per hour. So the relatively sedate Gulf Stream will never match the energy potential of the Pentland Firth if we only put a few turbines in the sea. But the Gulf Stream is wide enough to allow the installation of thousands of slow-moving devices; it also has the huge advantage that the speed is reasonably consistent throughout the year, unlike with tides. One researcher has calculated that the Gulf Stream ought to be able to provide a third of the electricity needs of Florida.

Nobody doubts that it is possible to build effective turbines in the Gulf Stream, since the waters are much less fierce than those off northern Scotland. The calmer water speeds mean that the turbine blades can be much bigger and more like windmills. The issue is money: given the low speed of the current, will it be possible to build an underwater windmill that can cover its costs? It all depends on the price of electricity in Florida. Other ocean currents, such as the relatively fast-flowing Kuroshio off Taiwan, could also be used for conversion into electricity at places where the worldwide oceanic conveyor belt, as it is sometimes known, comes close to populated coasts. Unfortunately, there are not many places that meet these criteria.

What about the wider environmental effects of slowing ocean currents by placing turbines in their way? Should people be worried in Britain, Norway and the rest of north-west Europe, where potentially bitter winters are kept several degrees warmer by the northern extension of the Gulf Stream? I think not, because proponents of Gulf Stream energy collection only intend to capture about a thousandth of the kinetic energy of the current.

OTEC

The final potential source of energy in the oceans is altogether different from the ones discussed so far. The idea is to exploit not the movement of water, but the difference in temperature between the warm surface waters and the colder depths. In parts of the central tropics, this temperature gradient is above 20°C, which in theory means these waters might warrant the installation of power stations based on the principles of the heat pump. The approach is known as ocean thermal energy conversion, or OTEC.

In a closed-circuit OTEC plant, the hot surface water passes through a heat exchanger, causing a low-boiling-point liquid, such as pressurised ammonia, to turn into gas. The expanding gas drives a turbine, thereby generating electricity, before exchanging heat with the cold water and condensing back into a liquid. And so on.

As with many other technologies discussed in this book, the basic idea is not new: scientists originally worked out how to turn a temperature difference into electricity over a hundred years ago. Several attempts to build a working generator have been attempted over the decades but the availability of cheap fossil fuels has always disrupted the experiments. Government research and development money dries up a few months after the oil price begins one of its periodic slips.

Even with higher fossil fuel prices and fears about climate change, however, OTEC is unlikely to prove a key technology in the coming decade. The main problem is that the ocean's surface temperature cannot rise above about 31°C. Higher than this level, the energy lost through the evaporation of water cancels out the energy gained from the sun. Since very few places near coasts have deep water at less than 4°C, the maximum temperature gradient is about 27°C, which is too small to create a very efficient heat engine. The best system might only capture a few per cent of the energy gradient, and will require significant amounts of power to run itself. Although it is perfectly possible to envisage a small positive energy balance, few but the band of enthusiasts backing the technology have much confidence that it will ever produce economical electric power. This is particularly true since this technology will only work in the tropics, where concentrated solar energy may become a cheaply available alternative.

Oceans as a whole, though, possess enormous untapped potential for low-carbon energy creation. Although it is still early days, I suspect that the best bets are wave collectors using the same principles as the Pelamis devices operating off Portugal and huge steel tidal-current collectors such as those produced by Clean Current and MCT. The manufacturing of such devices is relatively simple, and producers are unlikely to face any pressing shortages of raw materials. That means, once the devices are proven, there needn't be any huge delay before they're installed in their hundreds of thousands in oceans around the world.

Combined heat and power

Fuels cells and district heating

Most electricity generation today is inherently wasteful. Old coal-fired plants turn only about a third of the energy in their fuel into electricity. Even the best new gas plants struggle to reach 60 per cent. The rest of the energy becomes heat, which is treated as a waste product and frittered away in a cooling tower. Just down the road, thousands of people create yet more carbon emissions by burning gas to heat their homes. It doesn't really add up – environmentally or economically. A rational system would use the heat created in power stations to replace central heating boilers.

We have two obvious ways of avoiding this enormous waste of heat. We can switch to small power stations close to homes or offices and pipe the 'waste' heat to where it is needed. Or, on an even smaller scale, we can install microgenerators in our houses and places of work, making electricity precisely where and when we need it, and using the accompanying heat for room and water heating (or even for cooling, via a clever process known as absorption chilling). These two very different approaches both go by the broad name of 'combined heat and power', usually abbreviated to CHP.

Everybody agrees that CHP is a good idea: the towers of steam billowing out of large power stations are an increasingly obvious symbol of our profligacy with the world's scarce energy resources. But despite the attractiveness of the basic concept, CHP has struggled to grow in most countries. Using the waste heat from a small district electricity plant requires a network of underground pipes to take hot water to local homes and offices, an up-front cost that has provide a barrier during an era of cheap fossil fuels. One entrepreneur I talked to told me that it would cost £100 a metre to install the insulated pipework for a heating scheme in a European urban centre. Tiny heat-and-power units in individual buildings get around this problem, but unfortunately electricity generation on such a small scale has been very inefficient up until now, which means limited savings of carbon dioxide. Although CHP does make economic sense for some industrial processes in which the factory needs both heat and electricity, the costs have meant that take-up has not been particularly fast in recent years.

Thankfully, the obstacles to more widespread use of CHP are gradually disappearing. High fuel costs are making the installation costs of district heating pipes look more reasonable, and the efficiency of micro-CHP units is increasing. This chapter focuses on the two most interesting prospects for taking CHP forward: fuel cells powered by hydrogen created from renewable sources for individual buildings and district plants powered by wood and other biomass, again with a low carbon cost. Both approaches offer heat and electricity with minimal waste of heat and no use of fossil fuels.

Fuel cells

A fuel cell is effectively a battery. One side of the cell has a positive charge and the other has a negative charge. The two sides are separated by a semi-porous material called an electrolyte that allows electrically charged atoms to flow through. When an external wire connects the two sides, a current will flow around the entire circuit, just as with a battery. But unlike a standard disposable battery, the electric power of a fuel cell can be continuously topped up by the addition of more fuel, usually hydrogen at one side and oxygen at the other. For as long as there is fuel in the cell, a chemical reaction strips the positive charge from oxygen atoms and the negative charge from hydrogen, resulting in a reliable and consistent flow of electricity.

Getting oxygen into the cell is easy. The gas makes up a fifth of the atmosphere, so fuel cells simply feed ordinary air to the positive pole of the battery. Hydrogen is trickier. The pure form of this very light gas does not occur freely at ground level because it escapes upwards to the ozone layer, where it then reacts to form water and oxygen. (Loose hydrogen is therefore an ozone-depleting chemical.)

One option for production of pure hydrogen is to create it on an industrial scale by splitting water or hydrocarbons and storing the resulting gas in tanks that can be hooked up to fuel cells, where the gas will be efficiently and safely used. Despite its reputation, hydrogen is not particularly flammable or explosive.

Alternatively, the hydrogen can be made in the fuel cell itself. Perhaps the simplest way to do this is to use methane, the main component of natural gas, as the fuel. When heated to a very high temperature in the presence of steam, methane separates into its constituent elements: hydrogen and carbon. The carbon atoms combine with the oxygen in the water molecules in the steam to create carbon monoxide. (Usually known as 'steam reforming', this is the same process as discussed in the chapter on carbon capture and storage.) This leaves pure hydrogen gas. Inside the fuel cell, the hydrogen atoms then separate into their constituent parts: protons and electrons. The electrons, unable to travel through the innards of the cell, flow around the external circuit, providing electricity. The oxygen and hydrogen atoms eventually combine to form water, which is one of the two waste products of the process. The other is carbon dioxide, which is made from the oxygen and the carbon monoxide produced from the steam reforming of the original methane feedstock.

The voltage in an individual fuel cell is small – as little as 1 volt, which is less than that of an AA battery. So to make useful cells capable of driving large machines, many tiny cells are wired up together into a power pack. This means that fuel cells can vary in size from small devices for powering laptop computers or mobile phones to machines the size of several shipping containers capable of providing large buildings with all their electric power.

Like several other technologies in this book, fuel cells have been around for a long time but still have not achieved their full potential. The first cells were created in about 1842 by Sir William Grove, a scientist and lawyer originally from Swansea in South Wales. Intermittent attempts to create a commercial use for the technology followed, and in the 1950s several businesses tried to develop fuel cells to power vehicles or satellites. As with solar photovoltaic cells, however, the early promise shown in space missions proved difficult to translate into wider commercial success. For at least the last twenty years, several dogged manufacturers, such as Ballard Power in Canada, have been trying to build fuel cells that successfully compete with other energy sources, usually focusing on vehicles such as city buses. Progress has been painfully slow, for although the technology is well understood, it has been difficult to deliver a powerful fuel cell at a price that can compete with a standard internal combustion engine. Because some types of fuel cell, including Ballard's, use catalysts made from rare materials such as platinum, rapid upward shifts in metals prices have also impeded their development.

The most promising type of fuel cell

Several types of fuel cell are in production today, and they vary in important ways. Some need pure hydrogen as a fuel while others create their own from gases or liquids. Some operate at low temperatures while others work at the high temperatures necessary to split fuels into hydrogen and other chemical elements. Electrical efficiencies vary dramatically between the various types, but the best cells can now turn more than half of the usable energy in natural gas into electricity.

Ceramic Fuel Cells in Melbourne, Australia, is one of several businesses making good progress in constructing fuel cells for generating electricity and heat on a domestic scale. The company's cells are fuelled by the conventional domestic gas supply (which is largely methane) and employ one of the most promising fuel cell technologies, usually called the 'solid-oxide' approach. This description refers to the substance, made from the ceramic-like compound zirconia, that functions as the electrically porous centre of the cell. This technology uses very high temperatures, about 700°C, and it takes some time to get started, but does not need the extremely expensive platinum catalysts that are used in other types of fuel cell.

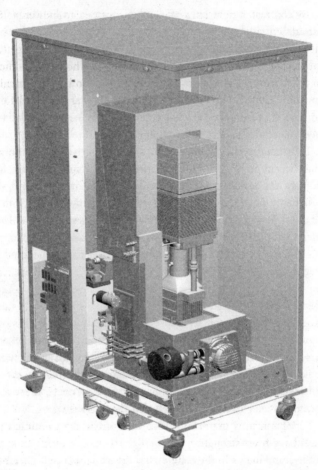

A drawing of a Ceramic Fuel Cells unit

Ceramic Fuel Cells' solid-oxide home power plants generate up to 2 kilowatts of electricity. The fuel cell system is almost 60 per cent efficient, meaning that it generates over 2 kilowatt-hours of electricity from natural gas that would produce 4 kilowatt-hours of heat if it were simply burnt. However, most of the rest of the energy produced by the fuel cell can be used for home heating, meaning that the device may be able to capture 85 per cent of the total energy of the gas and put it to use in the home.

By contrast, a new, ultra-efficient, large-scale gas power station can also turn about 60 per cent of the energy value of gas into electricity, but will suffer 5–10 per cent transmission loss in getting the electricity from the power station to users in homes and offices. In other words, the best fossil fuel plants are no better at turning gas into usable home electricity than one of Ceramic Fuel Cells' domestic-scale units. And when you factor in the useful heat generated, the domestic unit looks better overall.

In principle, solid-oxide fuel cells are very effective competitors to mainstream power generation. But, as the long-suffering investors in fuel cell companies are eager to explain, the entrepreneurs still have many hurdles to overcome. For one thing, there's size. Domestic fuel cells are still bigger than most central heating boilers. Even after a major effort to reduce the size of its domestic cell, Ceramic Fuel Cells' prototype is still the same size as a washing machine.

Then there's cost. Most of the companies in the fuel cell industry are coy about the price of their units, and Ceramic is no exception. We can safely assume that its product is still substantially more expensive than conventional central heating apparatus, although manufacturers are all promising continued sharp cost reductions. To get mass acceptance of the technology, solid-oxide fuel cells need to be priced to deliver electricity competitively. In the case of fuel cells, the industry thinks that means getting to below $2 per watt of continuous electricity output, or $4,000 for a 2-kilowatt cell.

Perhaps just as importantly, manufacturers like Ceramic Fuel Cells have been struggling to improve the length of life of some of the components in the cell. Most solid-oxide fuel cells currently only last three or four years before some components will need replacement or refurbishment. Customers would typically expect at least a ten-year operating life, and perhaps even longer, for a domestic boiler.

Another issue is that the attractiveness of domestic fuels cells will depend to a significant extent on the prices that homeowners can get for electricity fed into the grid. The first home CHP units were installed ten or more years ago. They used conventional internal combustion engines, not fuel cells, to generate heat from gas. The electricity generated came as a bonus. When the heat was not needed, such as in the summer or when the family was away, these CHP boilers were turned off, and the home took its electricity from the grid. Ceramic Fuel Cells has made many interesting innovations, but its key insight is that because its product can deliver an electricity output of well over 50 per cent of the energy value of the gas used, it makes sense to operate the unit 24 hours a day, exporting the electricity to the grid when it is not being used in the home. The unit could therefore be kept on even when the home is unoccupied. If the house does not need the associated heat, it can simply be exhausted to the open air, as at a power station. Some European countries, including Germany and France, have regulations that discourage the use of fuel cells when the heat is not actually being used, creating an obstacle to rapid take-up of these innovative devices.

The typical home will only use the maximum electricity output for a small portion of the day. (The electricity use in a European home is about half a kilowatt, averaged over the 24-hour day, and nearly double this in North America.) So the CFCL unit will satisfy the demands of the household almost all the time and will be exporting electricity from the house into the local grid. The value of the unit to the householder thus depends largely on how much he or she gets paid for this exported power. In countries with a 'net metering' requirement, the electricity utility is obliged to pay the same amount for exported electricity as it charges for household consumption. Some US states and Canadian provinces oblige the electricity suppliers to operate such a scheme. This provides a good deal for the owner of the fuel cell, who makes money running the cell all the time and exporting the electricity into the local grid. At current gas and power prices, the cell's owner might make a profit of $1,000 a year from electricity sales, even if all the heat is evacuated to the outside air.

But the heat needn't be wasted. If it works as promised, the fuel cell will provide all the heat and hot water needed for a very well-insulated house during the winter, even in quite cold countries. A big 'passivhaus' (see the next chapter) could be heated by a single fuel cell. In less energy-efficient houses, an integrated high-efficiency condensing boiler will be needed to supplement the heat. During the summer, the fuel cell will simply heat hot water for washing, venting the spare heat to the outside. As equipment costs come down, it will become increasingly attractive for householders to buy fuel cells and run them as micropower stations.

However, Ceramic Fuel Cells does not actually plan to sell its fuel cells to homeowners. It will provide them to the electricity utilities, which will then lease them to individual homes. Although the first units for commercial sale will not be produced until at least 2010, the company has pre-orders for 50,000 units from a Dutch utility and other orders from Germany. Relatively high local energy prices mean the early market for its fuel cells will probably largely be in Europe, so Ceramic is building its first factory at Heinsberg in Germany. The critical ceramic component, zirconia, that sits between the positive and negative poles of the cell is made in a specialised plant in northern England.

Other types of fuel cells

While Ceramic Fuel Cells is focusing on domestic homes, other manufacturers aim to address the market for larger machines for apartment blocks, offices, supermarkets and hospitals. US companies such as FuelCell Energy in Connecticut are already delivering units that produce more than 1 megawatt, enough to power an office block or a small shopping mall. Although FuelCell Energy's products are not currently based on the relatively new solid-oxide technology, they are efficient, reliable and attractive to utility customers. Like their smaller cousins, these large units can transform about 50 per cent of the usable chemical power of a fuel into electric power as well as generate large amounts of usable heat, either for keeping the building warm or cooling it via an adsorption chiller in summer. A 1-megawatt device will take up relatively little space and can be sited adjacent to the building. There are few safety issues and, unlike the back-up diesel generators frequently used by hospitals and other large buildings, the fuel cells cause no local air pollution. FuelCell Energy uses 'molten carbonate' technology for its cells, although it is hedging its risks by also being an active participant in the US government's research programme to improve the solid-oxide approach. (One-megawatt solid-oxide fuel cells are

An illustration of a large FuelCell Energy heat-and-power unit

likely to be available by mid-2010 from companies such as the UK's Rolls-Royce.)

FuelCell Energy's most important customer is the leading independent Korean electricity company, POSCO Power, which has ordered dozens of megawatts of capacity for delivery over the next few years, including some individual units as large as 2.4 megawatts. An installation of this size could produce about as much electricity each year as two large wind turbines in a windy location but only takes a fraction of the space.

Korea is poor in indigenous sources of energy, and its government is actively backing the use of fuel cells for electricity generation. It is supporting them by means of guaranteed high prices for their electricity output, similar to the 'feed-in tariffs' widely used to support renewable generation in countries such as Germany. The feed-in rates for fuel cell electricity in Korea are currently twice or three times the typical wholesale price of power. This substantial price incentive (which will gently decline over the next few years as the technology matures) has meant that many of the early orders for large-scale fuel cells have come from this country. The total installation is, of course, still only a small fraction of the output of a large coal or gas power station. At about $3,000 per kilowatt of power output, FuelCell Energy's plants are still expensive, though they compare well, for example, with the price of wind energy just a few years ago.

Japan has used a different approach, focusing on domestic rather than commercial fuel cells. The Japanese government has actively supported the installation of smaller cells by offering a subsidy that rebates a large part of the cost of the unit. In the first few years of the scheme, each installation was given a grant of tens of thousands of pounds. The amount is reducing annually as the cost of the cells falls. Progress has been much slower than the optimists predicted: in 2003, Japan announced what now looks like an absurdly ambitious target of generating 4.5 per cent of all its electric power from fuel cells by 2010. But the generous subsidy scheme has helped to start active competition in Japan between the proponents of the different fuel cell technologies. Indeed, this chapter's prediction that solid-oxide fuel cells will win the day is almost certain to be tested first in Japan. Ceramic Fuel Cells' units will be imported into Japan by its local partner, a large central heating boiler manufacturer, while the electronics giant Panasonic is putting its efforts into a different technology, similar to that of Ballard of Canada. Ballard itself is offering its low-temperature and well-established proton-exchange membrane approach in partnership with a Japanese business.

Panasonic has published its expectations for the Japanese market. By about 2012, it expects its fuel cells to make financial sense for householders to purchase without government subsidy. By 2015, it is forecasting sales of 300,000 domestic units, a very large number but still less than 10 per cent of all heating systems installed annually in the country. The manufacturing cost is still predicted to be high, at almost £3,000 per kilowatt of electric power output. Panasonic is emphasising the relatively long life of its units, suggesting that its products will work well for up to ten years after installation. But this is still a shorter life than a conventional home heating boiler that burns gas to heat water, so consumer acceptance is not guaranteed.

What fuel cells can do for emissions

Emissions from the use of fossil fuel energy to heat, cool and power buildings are as much as half of the total greenhouse gas output in most developed countries. Fuel cells offer huge potential for slashing these emissions.

A solid-oxide fuel cell powered by natural gas may reduce the greenhouse gas output of a home by 30 per cent or more, a much larger reduction than seen in early domestic combined heat and power units using internal combustion systems. But even the best performing fuel cell would still leave the typical north European house burning sufficient gas, mostly for winter heating, to produce emissions of 4 tonnes of carbon dioxide. So why are solid-oxide fuel cells such an important potential advance in the move towards a low-carbon future?

The answer is that eventually we will use renewable fuels to run these cells, rather than natural gas. In fact, one of the many advantages of solid-oxide cells is that they can be fuelled by a whole range of hydrocarbons, including cellulosic ethanol, the second-generation biofuel discussed in Chapter 7. If we power fuel cells this way, we will be reducing net emissions to a very low level indeed, perhaps as little as 10–15 per cent of the impact of natural gas. Fuel cells will also be able to exploit fuels from other renewable sources such as methane from slurry heaps or from sewage treatment plants.

A further advantage of fuel cells is that the waste products will usually only be carbon dioxide and water vapour. The vapour will be condensed back into a liquid and can be made sufficiently pure for drinking. The carbon dioxide then forms 100 per cent of the waste stream, and therefore can relatively easily be sequestered. (As Chapter 8 shows, one of the most difficult tasks in a carbon capture and storage process is separating the stream of waste gases to create almost pure carbon dioxide.)

This raises a very interesting possibility. At some stage, we should be able to use fuel cells to construct a fuel cycle that is 'carbon-negative', that is, better than carbon neutral. The cell will use ethanol made from renewable energy crops, such as switchgrass or wood wastes, which have absorbed carbon dioxide from the atmosphere. If the carbon dioxide produced during the operation of the fuel cell can then be captured and permanently sequestered, the whole cycle could actually result in a net removal of greenhouse gases from the atmosphere. This beneficial outcome is not something that will happen in less than a decade. Difficult technical issues need to be resolved, such as how safely and cheaply to compress and liquefy carbon dioxide in a relatively small scale. Nevertheless, this may eventually become one of the cheapest ways of reducing existing atmospheric carbon dioxide levels.

Fuel cells have another environmental benefit. Unlike fossil fuel generating plants, they produce almost no emissions of other polluting gases such as oxides of nitrogen or sulphur compounds. (A solid-oxide fuel cell powered by natural gas will scrub the small amount of sulphur before the fuel is separated into carbon monoxide and hydrogen.) In countries with unacceptable urban air quality, usually partly caused by old or inefficient coal- or oil-burning power stations, the fuel cell offers the prospect of real alleviation of dangerous atmospheric pollution.

Nevertheless, we should not overestimate the attractiveness of replacing gas boilers with fuel cells in domestic homes. For one thing, there's the unsolved issue of space. A cell with an integrated top-up boiler requires lots of room in itself. If it were to be powered by ethanol, the homeowner would also need space for a storage tank. Per unit of energy, ethanol takes up about 50 per cent more space than fuel oil, so the tank would need to be very large or replenished frequently.

These are important obstacles to fuel cells becoming the dominant source of domestic heat and power, but large-scale cells for offices and factories do not suffer from the same disadvantages. Take data centres for example. These buildings, containing racks of computer servers connected to the internet, now use over 2 per cent of the electricity produced in the world. They're putting strain on electricity supply and in some places, such as the Thames valley just west of London, they've even had their power supply capped. Data centres need power both to run the thousands of servers in the building and to provide cooling to prevent the computers overheating. They have high electricity demands 24 hours a day and (with the exception of a few newer centres with fresh-air cooling) they need substantial amounts of energy for cooling almost all of the year.

A large fuel cell power plant attached to a data centre would be the perfect solution. The cell could provide the electricity for the servers, while the heat created could power adsorption chillers to cool the building. This would avoid virtually all the waste associated with centralised electricity generation and provide the data centre with secure and reliable power. Of course, the building will still be connected to the wider electricity grid so that in the event of the fuel cell failing, power will always be available.

Moreover, though homeowners might find it inconvenient to have a large ethanol tank, operators of large buildings will have no such problems. Indeed, they may well already store diesel fuel to provide back-up power to protect against power cuts and other interruptions in supply. So keeping a stock of liquid fuel will involve very few extra costs. For major electricity users such as data centres, supermarkets, hospitals and large schools, the benefits of using fuel cells powered by renewable liquid hydrocarbons such as ethanol are overwhelming. Probably installed and maintained by the local power utility rather than the building owner, they promise to provide reliable, genuinely low-carbon power at reasonable prices.

Generating companies will also see big advantages in having large commercial buildings powered by onsite fuel cells. Not only will it enable them to serve extra customers in areas of tight supply. It will also give them a substantial source of replacement power at times when renewable electricity supply into the grid is limited. A large fuel cell in a school, for example, would face only minimal onsite demand in the evening. So it could either throttle back production or work flat out and export the excess power to the local grid. The utility company could have full remote control over the cell and increase electricity output at times of general power shortage. To illustrate precisely this point, Ceramic Fuels Cells has already successfully demonstrated that it can remotely adjust the output of one of its micro units from a control room tens of thousands of miles away. A small number of electricity utilities have begun to conduct trials to test how fuel cells can be automatically employed to adjust their power output at times of peak demand. If domestic and industrial cells can be turned up to full output when the power companies are short of electricity, they might represent another important buffer that allows the grids to accommodate larger and larger amounts of intermittent power from the wind or sun.

Solid-oxide fuel cells do not respond immediately to a call for more power: they can take ten or twenty minutes to adjust the electricity output to what is required. But this characteristic makes them perfect at matching the highly predictable daily swings in tidal power or the likely variations in solar energy. Concentrated solar power will give us daytime electricity from the Sahara. As the power of the North African sun ebbs away at the end of the day, fuel cells can be gradually ramped up to full output. In combination with other technologies that can provide power almost instantaneously, such as the pumped hydroelectric storage described in Chapter 1, large fuels cells offer a really substantial insurance as national electricity infrastructures become more exposed to the variations of renewable generation.

All told, then, fuel cells have enormous promise. They're still not competitive with large coal-fired power stations in terms of cost per unit of electricity generated, but we can be reasonably sure that much of our power, heating and even cooling will be eventually generated in fuel cell plants attached to homes, apartment blocks and commercial buildings. In time, these cells will be powered by renewable fuels, such as cellulose-based ethanol, making them even more environmentally attractive, and maybe even carbon-negative.

District heat and power

Fuel cells are scaling up to power and heat large buildings, but the other approach to combined heat and power seeks to supply entire urban areas. The story begins with district heating plants, which developed not because of fears over climate change, but because they offered households a relatively cheap way of obtaining heat in winter. Particularly in towns far from the mains gas network, a centrally located heating plant, often powered by local wood, provides a secure and inexpensive means of keeping the population warm. Hot water is distributed to homes in insulated pipes fanning out from heating plant. Users can adjust the flow of hot water through their radiators in the same way as they can change the settings in conventional central heating systems. The more hot water they use, the more they pay.

The prices charged vary from town to town, but most suppliers charge the equivalent of 3–4 pence per kilowatt-hour of heat. If you factor in the savings made by not having to install and maintain an expensive central heating system, this works out broadly competitive with traditional gas heating and cheaper than using oil. So most district heating systems around Europe offer householders reasonable value for money. Unlike solid-oxide fuel cells, they're already competitive with fossil fuels.

Increasingly, district heating plants are being used to generate electricity as well as heat. Like fuel cells, an efficient district plant can convert almost all the energy in a fuel into a mixture of power and heat. There is little waste of energy and very limited losses carrying the hot water or electricity to homes. Best of all, if the plant burns renewable fuels, such as local wood or municipal waste, it has only minimal greenhouse gas emissions.

Fuel cells and district heating plants have much in common, but they will be used in very different ways. Fuel cells are exciting because, thanks to recent advances, they can convert over half the energy value of gas into electric power. They can be put in a house or hospital and the secondary output, heat, is essentially free. If the heat (or cooling power) is used, then all well and good. If it is not, then little is lost. Ideally, the fuel cell runs all the time, sending electricity to the wider grid when its output is not required onsite.

District heating plants are different. They are not very efficient at converting fuels into electricity, perhaps providing only 20 per cent of their energy output in the form of power, with the other 80 per cent being heat. This means district heat and power plants are usually operational only when they are able to sell heat – or the equivalent amount of cooling power. Many future plants should get around this problem thanks to gasification technology. First, the fuel is heated to about 700°C in a low-oxygen environment, just like the charcoal-making process discussed in Chapter 9. The heat drives off hydrogen and carbon monoxide, both of which can be used as fuels to drive turbines or heat water. Once this process is complete, the remaining charcoal can be burnt. One of the first large-scale biomass gasification plants in Europe has been successfully operating in the Austrian forest town of Güssing for several years, producing 2 megawatts of electricity and 4.5 megawatts of heat for the district heating system. The ratio of the output of electricity to heat may make it economic to run the plant even when the heat is not needed.

Some countries use district heating extensively, others barely at all. The UK and the US, for example, rely largely on central heating boilers in each home, whereas Canada has about 150 working district heating schemes, mostly serving city centre buildings. However, the Canadian town of Charlottetown on Prince Edward Island started its own local hot water system in 1986, burning wood chips from otherwise unusable local timber. The system also generates electricity using a steam turbine. The network provides heat to eighty town buildings, including university, hospital and government offices as well as apartment blocks and a small number of private homes. The heating systems in North America are generally small in extent and cover few homes.

In Denmark almost two-thirds of the population get their hot water and heat from over 400 district heating systems. Some of these are cooperatively owned ventures, serving a few hundred households, while the biggest provide heat to 100,000 homes or more. About three-quarters of the plants in Denmark also generate electricity. A fuel is burnt, steam is generated, and some of this steam drives electric turbines, while most of the hot water is separated off for heating. The district heating trade association claims that more than 90 per cent of the energy in fuel is typically converted to usable heat or electricity. Most Danish plants are still fuelled by gas or other fossil fuels, but about 40 per cent of the heat produced is carbon dioxide free, because the plant has burnt renewable wood or domestic waste. (If the waste had been buried instead of burned, it would have decomposed in a landfill site producing methane and other greenhouse gases, some of which would have escaped to the atmosphere. Since methane is a worse greenhouse gas than carbon dioxide, there are clear climate change benefits to using waste food in this way.)

The carbon savings from using district heating in Denmark are said to be substantial. An independent trust which works to improve energy efficiency says that the typical Danish district heating scheme produces heat with only a quarter of the carbon footprint of heating a home by electricity. A householder moving from electricity to district heating will generally consume more heat because it is cheaper. But the increase in consumption is typically only about 15 per cent, so it wipes out only a small fraction of the carbon savings.

Over the Atlantic, Canadian company Nexterra is pushing gasification technology for large-scale plants designed to burn forest wastes. According to one estimate, almost half of all material taken out of forests is not used to make timber or pulp, and is therefore available as fuel for decentralised power stations. One project sees Nexterra's gasification technology, praised by industry insiders for being simple and reliable, installed as the heating plant for a new urban community in the port area of Victoria, British Columbia. This innovative technology helped win the new community an award from the Clinton Climate Initiative in July 2009 as one of just 16 'Climate Positive Developments' worldwide, recognising that the Nexterra plant provides a net surplus of energy for export to the wider area. The plant is flexible enough to provide just heat, just electricity, or a mixture of the two. Nexterra is also going to provide a biomass gasification plant for heating the campus of the US government's premier energy research centre at the Oak Ridge National Laboratory in Tennessee. Another proposal takes gasification technology to fifteen communities in Canada's western interior where the wood stock has been seriously affected by pine beetles that kill the trees and reduce the value of the timber. The network of wood gasifiers would each generate up to 10 megawatts, providing power for thousands of homes in communities whose economic viability has been seriously affected by the destruction caused by the beetles.

How big a difference could wood-powered CHP plants make

The combined heat and power plant in Borås, Sweden

to our low-carbon future? The potential is huge, but there is one key question: the availability of land to grow trees. Take Sweden, a country where many municipalities use district heat and power plants. For example, the plant in Borås, near Gothenburg, generates about 25,000 homes and 2,000 offices with most of their heating needs and much of their electricity. To do this, it uses about 270,000 tonnes of wood chips a year, or approximately 10 tonnes per customer. Sweden as a nation produces about 30 million tonnes a year of wood from its forests, or only about 6 tonnes per household. Therefore even if *all* the Swedish wood harvest was used for district heat and electricity, there wouldn't be enough to keep everyone in the country warm. If one of the world's major exporters of forest products does not produce enough energy in its timber, surely very few other countries will have enough raw material either.

This interpretation is too harsh. In truth, the amount of wood taken from Sweden's forests is governed by world demand for paper and timber, not by a shortage of woodland. Sweden boasts several billion tonnes of standing trees, compared with an annual need of less than 50 million tonnes to keep its entire population warm with district heating. Indeed, the European continent as a whole has a billion hectares of forest land and if 20 per cent of this land were to be converted to fast growing species of trees, such as willows in damp temperate countries, it would provide enough energy to heat all European homes at today's patterns of consumption. In a densely populated country like the UK, with a relatively small area of forest, about 10 per cent of the entire land area would need to be given over to growing wood for fuel, though this figure could be reduced substantially by improving the insulation of existing buildings, as discussed in Chapter 5. In the case of Canada, the numbers are even more compelling. About a third of the country's landmass – about 300m hectares – is thickly forested and about half this land is accessible. Yet less than 1m hectares is harvested each year. The whole of Canada's heat needs and a large fraction of those of the US could be met by sustainable use of northern forests.

However, the problem is that heating plants aren't the only source of new demand for wood. Two other chapters of this book – those on cellulosic ethanol and biochar – also focus on prospective uses for woody matter grown on the world's limited supply of reasonably fertile and well-watered land. The crucial question is how best to balance the primary requirement to use land for growing food for an increasing population, on the one hand, and for harvesting energy crops on the other. The striking food price rises of 2007/8 provided a sharp reminder of the inherent conflicts between food production and the world's need for energy as corn crops were diverted to ethanol production for motor cars.

In Chapter 10 I propose that one answer to this conflict may be to renew our focus on improving the huge areas of the world with degraded soils on which virtually nothing grows. Any real solution to climate change must involve the restoring of soil health and the rolling back of desertification, probably using massive tree-planting schemes such as those employed in western China. Increasing the amount of usable land on the earth's surface will help us meet the challenge of growing more woody biomass and feeding 3 billion more people than the globe sustains at present.

Super-efficient homes

Passivhaus and eco-renovations

In 1991, a terrace of four new homes was completed in Darmstadt, Germany. From the outside they looked much like similar houses completed in Germany at the time, but there was a crucial difference. None of the homes featured a conventional central heating system. This wasn't an oversight or an exercise in promoting ascetic living. These houses, designed by a group of Swedish academic architects led by a quietly spoken German engineer called Wolfgang Feist, were the first passivhaus buildings, which means they were so well insulated and cleverly designed that they didn't need a full central heating system, nor, indeed, an air-conditioner for the summer.

Heating and cooling don't generally get quite the same media attention in the climate-change debate as cars and electricity-hungry gadgets. But they should. If you added up the emissions of all the world's gas and oil boilers, coal fires, electric heaters and air-conditioning units, then you'd probably find that managing the temperature of buildings – either through heating or air-conditioning – is the world's single most climate-damaging activity.

A large slice of heating and cooling emissions are created needlessly, since almost all homes and offices are a very long way from passivhaus levels of insulation and intelligent thermal design. Much of the hot and cold air created by boilers and air-conditioning units is haemorrhaged through leaky walls, windows, floors and roofs. This is a huge economic waste as well as an environmental problem, with

The original terrace of passivhaus homes in Darmstadt

soaring energy costs making it more and more expensive for house-holders and businesses to maintain a comfortable temperature.

Over the last twenty years most developed countries have introduced regulations that demand increasingly high insulation standards when new buildings are constructed. But governments around the world have been surprisingly slow to push for better energy efficiency in existing buildings. This is a mistake: in a typical

European country the number of new homes constructed each year may be less than 1 per cent of the number of houses in the existing stock and even in the US it is less than 2 per cent. We face, therefore, a far more urgent need to get homeowners to refurbish their existing homes than to impose higher and higher insulation standards on new building. Unfortunately it is easier and less intrusive to impose regulations on construction companies than to institute a massive programme of eco-renovation. With a few exceptions, governments are avoiding dealing with the obvious need to make substantial improvements in older buildings.

Nevertheless, there is substantial reason for optimism. Unlike some of the technological innovations in this book, many domestic energy efficiency measures make financial sense even in the short term. We don't need further technological improvements or a high tax on carbon emissions. For example, at today's energy prices, it is often sensible for the homeowner to very significantly improve the insulation standards of the home, especially in the countries where winter temperatures are very low. International experience, particularly in Germany, is that a wide-ranging programme of information provision, exhortation, subsidy and cheap loans can successfully push landlords and homeowners into taking action. Implemented enthusiastically, savings in energy use can exceed 60 per cent and sometimes reach 80 per cent. One important eco-renovation in Austria of a 1950s apartment block cut energy use by 90 per cent.

There really isn't any alternative to improving the efficiency of existing houses. We cannot simply tear down the hundreds of millions of leaky homes that exist today – we need to find ways of reducing energy use while leaving the fabric of the building intact. There will be aesthetic issues: the most effective way of reducing energy use is to introduce a thick layer of insulation on the exterior of the walls of houses. This may change the appearance of the building and arouse strong local opposition. Insulation invisibly attached to the inside of the house can be nearly as effective but it does reduce the size of the rooms. Combined with better windows, floor and loft insulation, and good central heating boilers, eco-refurbishments could cut carbon dioxide emissions by a substantial percentage.

Before looking at renovations, however, this chapter explores the ideas behind the passivhaus movement. Although only a few thousand houses have been built to passivhaus standards, these buildings have shown that huge reductions in energy use can be designed into all houses, not just expensive eco-homes. Good construction techniques and ruthless attention to detail matter as much as the choice of insulating material. Passivhaus thinking has become embedded in new building activity in several different parts of the world, but also needs to inform the massive programme of eco-renovation that the developed world urgently needs.

Passivhaus

The passivhaus idea is simple. A house insulated to the highest standards does not actually need a central heating system. Even in the depths of winter, it can be kept warm by capturing energy from the sun, and from the heat given off by the people and electrical appliances it contains. On the coldest days in high latitudes, the building may need a top-up from an electric radiator but even in cold countries a well-built house can remain comfortably warm during winter. In hot climates, passivhaus construction can help to dramatically reduce the need for electric air-conditioning.

The first houses built to the passivhaus ideal were constructed in Germany almost twenty years ago, so the idea has taken a long time to blossom. Even now, there are probably fewer than 15,000 certified passivhaus homes around the world and most of them are in Germany and Austria.

The slow take-up is surprising. A properly built passivhaus dwelling should have energy consumption for heating of less than 10 kilowatt-hours a year for each square metre of floor area. That's around sixteen times less than the average home in the UK, and eighteen times less than the average in Germany, with its colder winter temperatures. Getting from the levels of energy use typically seen today in Germany and the UK down to the levels that can be achieved by full eco-refurbishment will save several tonnes of carbon dioxide per house, or at least as much as completely eliminating the greenhouse gas emissions from the household's car.

The intellectual force behind the passivhaus standard remains the German engineer who build the first Darmstadt house, Wolfgang

Feist. He went on to found the Passivhaus Institute, a body dedicated to setting the standards for energy efficiency in home construction. It is probably significant that Feist is an engineer by training, and not primarily an architect. Indeed, listening to one of his talks, you quickly understand that reducing the carbon emissions from housing is largely an engineering challenge and has relatively little to do with architecture, at least as it is conventionally understood. Passivhaus homes do not need to look any different from the prevailing architecture of the area, and they certainly don't have to be small or strangely shaped. How a house looks isn't important: to get passivhaus certification is simply a matter of meeting energy use requirements. Neither does a passivhaus have to be high-tech, full of steel, concrete and granite and controlled by sophisticated electronics. And although a passivhaus will often use solar collectors for heating hot water, there is no need for expensive photovoltaic panels or domestic wind turbines.

According to Wolfgang Feist, achieving energy efficiency in new housing simply requires the builder or renovator to focus on five key principles: excellent wall insulation; small, high-quality windows; air tightness; a lack of 'bridges' that conduct cold into the house from the outside air; and a ventilation system that brings fresh air into the house and preheats it using warm, stale air extracted from the main rooms.

Putting all these elements together is not a simple matter and can be expensive if done without thought. But there is no magic or unusual technology involved. Let's look at each principle in turn.

Wall insulation

Walls are the main source of heat loss in most homes. Although huge amounts of energy can be lost through the roof, most houses have sufficient loft insulation to reduce the outflow of heat (though almost all would benefit from another layer). Walls are a more difficult problem. To meet passivhaus standards, a home constructed from bricks will need a thick layer of insulation either on the exterior wall, in which case the facing brick will be invisible, or on the inside walls of the house. This insulation will need to be around 40 centimetres (almost 16 inches) thick, significantly reducing the internal dimensions of a room, if applied internally, or adding to the bulk of the house if used on the exterior walls. This insulation will usually be made from expanded plastics, but a wide variety of alternatives are available.

Windows

Even very well-insulated windows let in more cold than a wall, so an energy-efficient house needs to have a relatively small percentage of its surface area given over to glass. This doesn't need to make the house dark; some of the lightest homes I have ever seen have just been constructed on a large eco-estate in Milton Keynes, north of London. Although they weren't built to passivhaus standards, they are better insulated than any other mass-produced UK houses. The windows were intelligently positioned to capture as much light as possible, particularly in winter.

The passivhaus approach is to face all large windows to the south (or to the north in the southern hemisphere). This maximises not only the incoming light but also the heat from the sun caught by the house in the winter months. The passivhaus standard looks for 40 per cent of the total winter heating need to be met from the sun's heat entering the house through window glass. If a shade is put above the south-facing windows, the high angle of the summer sun means that relatively little unwelcome heat is captured in the hottest months of the year. When the sun is high in the sky, it is most important to have a well-insulated loft that blocks the heat from entering the house through the roof.

Dr Feist points out that perhaps only 70 per cent of a window is glass. The rest is the frame and the fittings. Energy losses from these elements can be far worse than from the glass. So considerable effort has gone into designing frames that do not leak heat. This sounds a simple task, but the engineering is actually very complex. The photograph opposite shows some of the complexity in one of the many innovative window designs that meet the insulation requirements for a passivhaus. Such windows are usually triple glazed, with inert gas inside and glass surfaces that reflect heat back into the house. Even these products probably emit five or six times as much heat as a really well-insulated wall, so they cannot be too large.

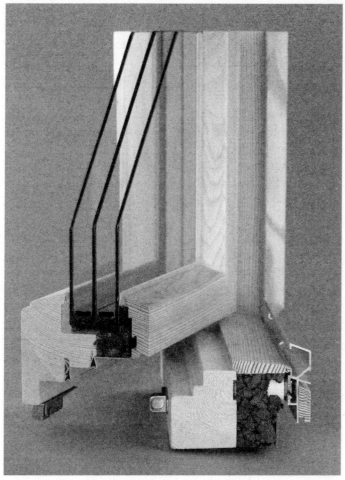

*A passivhaus-certified window by Optiwin, with three
layers of glass and various thermal barriers*

Feist gives some illuminating figures for the impact of good
window design on levels of internal comfort in the winter. A well-
made triple-glazed argon-filled window will feel warm even when
temperatures are well below freezing outside. He says that the best
examples can keep the temperature at 18°C on the inside of the glass,
compared to just 5°C for a traditional double-glazed window.

The difference this makes to the feeling of comfort in the room is very marked. A room with an air temperature of 19°C but with warm windows will often feel more comfortable than a room at 21°C but with cold temperatures at the window glass. The reason is that the warm human body is losing radiant heat to the cold window. Further discomfort is caused if a person is standing or sitting sideways to the window. He or she will be giving up more radiant energy on one side of the body than the other, which tends to feel uncomfortable.

'Bridges' that conduct cold into the house

Conventional construction techniques often allow very conductive materials, such as metals and concrete, to provide a bridge between the cold outside air and the inside of a house. Even the best-built homes often lose significant amounts of heat in this way because of poor design and carelessness during construction. Passivhaus homes avoid the problem by the use of carefully prefabricated components and vigilance during the building process. Factory prefabrication of houses – still unusual in some parts of the world and anathema in places like the UK – helps reduce heat losses because components are made to more accurate specifications in clean and dry conditions in a factory. Passivhaus homes don't necessarily have to be factory made and then assembled on the building plot, but this is the easiest way to achieve the standards required.

Airtightness

Even the best conventional homes lose huge amounts of warm air through cracks, airbricks, poor door seals and other routes. New construction techniques vary around the world but few builders anywhere understand how important airtightness is to the overall energy consumption of the home. One UK government body has said that today's 'best practice' still produces over three times the level of air loss required in the passivhaus standard. In a typical new house perhaps 30 or 40 per cent of the heating requirement arises because of the ingress of cold air through gaps created accidentally during construction. To get a passivhaus certificate, the building needs to pass a test in which the air pressure in the house is increased and the rate of air loss to the outside world is measured. To pass the test, the house must lose less than 60 per cent of the volume of air in the house per hour. By comparison, an older conventionally built house will often have ten or twenty times this rate of leakage, particularly in windy locations. Those who cannot believe this figure should sprinkle a fine powder, such as talc, at the corners of an older room on a windy day and watch the moving air blowing it around.

Creating a fairly airtight new home is difficult, even if the components are factory made and fit together very tightly. Home building tends not to be a business that attracts finicky perfectionists used to working in fractions of millimetres, but that is what is required to get a passivhaus building to achieve its full potential.

The ventilation system

Because a passivhaus home is so airtight, it needs to have a ventilation system that brings fresh air into the house and extracts the stale air. Otherwise the occupants would suffer from excess carbon dioxide, which builds up as a result of human respiration. The house would also suffer from pollution from other sources such as the unpleasant chemicals given off by most paints. But a simple air extraction system using fans would be no better than a leaky house: warm air would leave and cold air would come in.

The passivhaus solution is the use of heat exchangers: cold air entering the house passes over ducts containing the warm (and humid) air leaving the building. This kind of ingenious heat recovery system can transfer a remarkable 80 per cent of the outgoing heat to the incoming air and also provides the ideal place to top up the heating, when required. Electric elements can heat the incoming air on the few days a year that this might be necessary. Typically the air in the whole house will be changed every couple of hours, ensuring abundant ventilation as well as excellent energy efficiency.

Passivhaus principles also work in hot countries. A thick and effective barrier of insulation combined with airtightness and forced ventilation can all work to keep the heat out of a house just as well as they ensure winter comfort in cold countries. Although few certified passivhaus homes have so far been built in the tropics, the passivhaus principles make perfect sense in such regions. The key difference between northern Germany and Australia might be the way that air is cooled before coming into the house. In a very hot country it could, for example, be passed along a duct running 2 metres under the garden surface. In the middle of the day, temperatures well below the soil surface are much lower than in the air.

The pioneering passivhaus homes in Darmstadt have been the subject of much research during the past two decades. One key finding is not only do residents have extremely low heat demand but that electricity use for appliances and lights is also well below German averages. In one sense this is not surprising: we might expect people who live in passivhaus homes to be interested in energy efficiency. And indeed, some researchers have suggested that a large part of the total energy savings arise because householders are highly motivated to run the house efficiently. Wolfgang Feist points out that the evidence tends to contradict this hypothesis: there is as much variation in energy use between different passivhaus homes as there is between conventional houses. Some passivhaus householders are relatively profligate in their use of energy and others are extremely careful. So the low average energy use does not result simply from the occupants neurotically trying to reduce their energy bills. The houses are light, so the need for electricity for lighting is low. The absence of a central heating system also reduces power use because of the lack of pumps and controls. Passivhaus standards really do radically diminish the average use of energy in all types of houses and in all temperature zones.

It is difficult to come up with many substantial disadvantages to the passivhaus approach. Perhaps surprisingly, for many people the most troubling effect of living in a passivhaus home is that most external noise disappears. If the windows are closed, the sounds of birdsong, light traffic, or children playing in the street are all absent. Another difference is that the air in the house tends to feel very dry. Air in homes is made fairly humid by human respiration and by water vapour from kettles, showers and cooking. In a passivhaus home the mechanical ventilation system extracts this wet air and replaces it with colder air from the outside. Cold air can hold very little water vapour compared with hotter air, and external air coming into the house at −5 °C will often be nearly saturated, containing as much water as it can possibly hold. The humidity level of this air falls when it is heated by the heat exchanger in the ventilation system. Air at 20°C can hold five times as much water vapour as air at −5 °C, so it is now only 20 per cent saturated with water vapour. This figure is called the 'relative' humidity level. Human beings are more sensitive to the relative humidity level (how much water vapour the air contains compared to the maximum it could contain at that temperature) than to the actual amount of water in the air. Most people feel comfortable with relative humidity levels over 40 per cent, so the passivhaus mechanical ventilation system will inevitably make the house feel dry when the temperature is very cold outside. This effect is exacerbated if the air is dusty. So the owners of passivhaus homes need to invest in good vacuum cleaners as well as humidifiers and large numbers of potted plants to maintain moisture levels at all times. Neither of these two disadvantages seem particularly off-putting.

The need for substantial extra blocks of internal insulation means that the rooms are very slightly smaller than would otherwise be the case, but the effect is marginal. In one house I looked at, the thicker walls had reduced the internal space by less than 3 per cent. That's barely noticeable in a large house and it's partly offset by the space saved by not having radiators on many of the internal walls.

How much do passivhaus homes cost and what are they like to live in? I spoke to a small housebuilder in the remote west of Ireland to get a view from a country that has only recently started to build them. Scandinavian Homes imports extremely well-insulated prefabricated housing from Sweden and decided four years ago to offer a passivhaus upgrade to its existing product line. It built its first demonstration house in 2005 and has been constructing passivhaus homes for sale since 2006. By a fortunate chance, the person who answered the telephone when I rang the office happened to be the company's first passivhaus customer. After her home was constructed, she eventually became an employee of the business. Miriam Green's family moved into their Galway home on the windy and wet western coast of Ireland in 2006. 'The first winter wasn't easy,' she said. 'We thought we'd get enough warmth

The first prototype passivhaus in Ireland

from the electric appliances and from body heat but we were wrong. We didn't realise that the rules allowed us to use some underfloor electric heating in the depths of winter as well. But our usage of heat generated by electricity never went above the passivhaus standard of 10 watts per square metre.' (This is less than 10 per cent of a typical Irish home.)

Miriam's house is large – almost double the average size of a European home – so it is not surprising that heat from the occupants and the electric appliances is not quite enough to keep the building warm on the coldest days. Another issue for houses in cloudy western Ireland is that the winter climate offers little solar warmth. Although temperatures don't often go much lower than freezing, the overcast days reduce the amount of solar energy coming in through the south-facing windows. The 2007–8 winter was particularly cloudy and she noticed the effect on the temperatures in the house. Miriam Green also commented on another problem: the roof windows of her home, imported from a large manufacturer in Scandinavia, are simply not well enough designed. They are triple glazed but the seals around the edge do not fit snugly and on the windiest days she can feel a draught. (A problem that chimes with Wolgang Feist's passionate focus on the improvement of window frames.) Miriam says, however, that the single-storey passivhaus homes that her company offers never suffer from this problem because they don't use roof windows.

The amount of electricity that Miriam's family uses reflects the high insulation standards. She and I easily calculated that, in the coldest month of the past year, the house must have used about 800 kilowatt-hours, of which well over half will probably have been used to run the home appliances. Heating demand was perhaps a tenth of a poorly insulated house of the same size, very much in line with passivhaus expectations. But electricity is a very expensive way of heating a house – perhaps four times the price of gas for an equivalent amount of energy – so the savings in cash for the Green family are not likely to be as great. And as an extremely ecologic-ally aware individual, Miriam was concerned that heating a house with electricity, albeit in relatively small amounts, was bad practice because of the carbon dioxide implications of power generation. The figure varies between countries depending on the type of fuel used to generate electricity but a kilowatt-hour of electricity usually produces two or three times as much carbon dioxide as a good gas boiler delivering the same amount of energy. So she is looking to install an innovative 'heat pump' to replace the electric underfloor heating. Heat pumps still need electricity for their power but a modern system will produce three units of warmth for every unit of electricity consumed.

The homes built by this small Irish company were already highly efficient and were engineered to be airtight even before they offered a passivhaus upgrade. The mainstream houses of Scandinavian Homes aren't representative of the average cost of constructing a house in Ireland, so it is not easy to work out the extra cost of build-ing to passivhaus standards. Nevertheless, in her role as employee of the firm, Miriam gave me some estimates. She said that she thought that the materials cost of a typical passivhaus bungalow was about 5 per cent higher than a conventional home from her company but a two-storey house might cost 15–20 per cent more. Most of this incremental cost arises because of the extra internal insulation. On the other hand, the lack of any need for a central heating boiler and room radiators means a big saving. All told, Miriam reckons that upgrading a very well-insulated house into the passivhaus category probably adds in the region of 10 per cent to the aggregate cost of materials and labour.

These figures are consistent with the estimates provided by Wolfgang Feist of the Passivhaus Institute. Based on German prices Dr Feist says that the total cost of building a passivhaus is about €15,000 greater than for a comparable home constructed to today's government-mandated insulation standards. Subtract the saving made by not having a central heating system and the incremental cost works out at about 8 per cent of the average construction costs of a new German house, a similar figure to Miriam Green's estimate. Not only will such a house provide comfort during the winters and hot summers, but the lower energy bills would probably pay back the investment in a dozen years or so. In other words, anybody wanting a new house would be very well advised to buy one built to passivhaus standards.

So why aren't more houses being built the passivhaus way? Miriam Green answers from her own experience. Her family found the prospect of a passivhaus home intimidating. 'It required a huge leap of faith,' she said. 'We were the first people in Ireland to commit to buying a passivhaus home and we took some convincing that the house would be warm.' But now, as an employee of a pioneering building company, she finds attitudes are changing fast. 'People in Ireland now understand what a passivhaus is,' she says. 'They're much easier to sell now.'

There's also another problem. The housebuilding industry simply isn't used to constructing energy-efficient homes. To start building to passivhaus standards will completely change the way large house-builders do business. Until the workers on a construction site have been fully trained, every single phase of the complicated process of building a house will have to be closely supervised. Many of the construction techniques that the industry has evolved over the last half century to build homes rapidly and at low cost with relatively unskilled labour are simply not compatible with the relentless attention to detail required by the passivhaus approach. I talked to two large housebuilding firms in the UK which had pioneered small developments built to high-insulation standards, though neither had yet constructed any housing to full passivhaus rules. Both told me that their first homes had cost well over 40 per cent more than comparable developments elsewhere. They acknowledged the cost increase arose because of inexperience, mistakes in design, and problems in sourcing the unusual materials. But there is a learning curve in housebuilding just as much as in the manufacturing of wind turbines, and as housebuilders get more experienced, costs will fall to the level of the small builders who have pioneered passivhaus construction.

Zero-carbon homes and eco-renovations

Interest in super-efficient homes is there, but low-carbon housing is really not taking off at the pace that it needs to if we are to see a substantial reduction in overall emissions from the housing stock. Even in Germany passivhaus construction accounts for only a small proportion of total building, although some people talk about getting it up to 20 per cent of all new buildings within a few years. The lack of progress means that countries like the UK are now using the law to force builders to start constructing really energy-efficient housing, even though customers ought to want them anyway because of the lower energy bills.

Many doubt whether the government's target will be met, but all new housing in the UK must be 'zero-carbon' by 2016. Increasingly stringent requirements apply as we move closer to this date. The UK target is astonishingly ambitious, not yet matched by any other country in the world. 'Zero-carbon' means that any fossil fuels used in the house – oil, electricity and gas – must be balanced by renewable energy generated on the same site, or very nearby. So where the passivhaus principle reduces heating use to perhaps 10 per cent of that in a conventional house, and may cut the electricity needed to power the lights and appliances in half, the UK rules for new construction will become far more demanding.

At first hearing, the UK approach sounds attractively purposeful. The government has set a huge challenge, but one that is theoretically possible to achieve. British construction companies have divided into those that think it will make new housing impossibly expensive and those who see the targets as an interesting way of stimulating a rather conservative industry into rapid change.

One of those more optimistic businesses is the Irish building materials firm, Kingspan. Its UK subsidiary has produced plans for a 'self-build' house – constructed by tradespeople managed by the purchaser of the new building – that is zero-carbon because it has many solar photovoltaic panels on the roof combined with very high levels of insulation and airtightness. A wood-burning boiler provides supplementary heat and because wood is classed as a renewable fuel, the carbon dioxide output is not counted in the 'zero-carbon' calculation. However, this house, called the Lighthouse, is extremely expensive. It costs about £180 per square foot to build, well over twice the construction cost of a less energy-efficient house built on the same site. (For comparison, some passivhaus buildings in Germany have been built for less than £100 per square foot.)

From the outside, the Lighthouse is extraordinarily attractive, looking like the spinnaker of an ocean-going yacht in full sail. Inside it seems a bit cramped, particular in the bedroom areas on the bottom floor of the house. Light levels around the building are good, despite the relatively small percentage of the wall area given over to windows. But whether the typical new home buyer in the UK is going to spend willingly an extra £100,000 to live in an ultra-low-carbon house is something I would strongly doubt.

Moreover, when it comes to carbon emissions, the simple fact is that this £100,000 could be much better spent. The law of diminishing returns applies. A zero-carbon house is always likely to be far more expensive than a house built to the passivhaus standard, but the incremental savings in carbon emissions are low. It would be much better to use £100,000 to renovate the oldest buildings and take them up to modern standards of insulation and airtightness. This would have many times more impact.

The Potton Lighthouse

I talked to Tim Fenn, who runs a building firm that focuses on energy-efficient renovation near Oxford. He said that a typical large and draughty Victorian house in Britain, owned by a family wanting to maintain the traditional external appearance of bricks and mortar, might consider two different types of improvement. Taken together, these improvements might cost £25,000 and save 75 per cent of the heating bill in an old house – and around 4 or 5 tonnes of carbon dioxide a year. If the house is being fully renovated, the first improvement would be to install sheets of insulation on the inside of the exterior walls. And, to improve airtightness and reduce heat loss, the renovator can install a thin insulation material faced with aluminium foil to block the flow of radiant heat and air from the room.

Heat moves from a hot place to a cold place by three mechanisms: radiation, convection and conduction. A layer of 1-centimetre-deep, foil-faced insulation does relatively little to stop conduction of heat, but is very effective at blocking infrared radiation. Experts disagree on just how much of the heat loss from a house is accounted for by radiation, but some claim it is as much as a half. Just adding reflective foil membranes to the inside of the main rooms in an eco-renovation might avoid a third to a half of the heat loss from an old house. It would also keep the house cooler in summer because the aluminium would help block solar radiation from entering the house.

The very poorly insulated British housing stock would benefit from substantial renovations of the type that Tim Fenn identifies. It would be even better to add insulation to the *outside* of old buildings, but, as the building industry acknowledges, the British are very fond of keeping the external appearance of buildings unchanged. After all, houses from the nineteenth century and earlier are a vital part of Britain's urban landscape. So in the leaky, cold and damp Victorian buildings of the nation's towns and cities, where one can often see the world outside through cracks in the lovingly restored pine front door, the insulation will have to be installed inside, slightly decreasing the size of the rooms.

In areas where the external appearance can be changed, the best approach is to resurface the outside of buildings either with 5–10 centimetres of plastic cladding, or with 1–2 centimetres of reflective foil-faced insulation of the type described by Tim Fenn. The surface can then be rendered with a coating to provide an appropriate colour or finish. Old blocks of flats are particularly suitable for this type of treatment. It doesn't produce a passivhaus level of insulation, but it will very substantially improve the heat losses from older buildings. It will also reduce the condensation problems that arise when the water vapour in the heated internal air meets the poorly insulated external wall. Construction companies around the world are reluctant to acknowledge this, but these are lessons learnt by the passivhaus movement in Germany and elsewhere. The importance of extraordinary attention to detail to ensure complete airtightness and an absence of 'bridges' that conduct heat to the outside world is increasingly clear. Extreme care needs to be taken both in designing refurbishments and in actually carrying out the work. In most countries, the home construction industry is still some way from fully absorbing this point.

Great Britain's housing stock also includes a large number of homes (more than half of all houses) using two parallel walls of brick or concrete block separated by an air-filled cavity. If this cavity is pumped full of insulation, the energy efficiency of the building improves substantially, and at a cost of just a few hundred pounds. This outlay is usually repaid in lower fuel bills within two or three years. Nevertheless, even at a time of great concern about high fuel bills, most householders who could carry out this simple eco-refurbishment have failed to do so.

These are the easiest ways to significantly improve the amount of heat lost from British homes. But compared to its commitment to improving the standards of new buildings, the UK government has few measures in place to actively encourage householders to renovate their houses to improve energy efficiency. There's a limited programme to increase the numbers of homes installing wall-cavity insulation, but little else. The result is that the typical energy efficiency of British homes is barely improving.

The German administration is very different. A recent government statement said the renovation of existing buildings is 'a central element in Germany's national climate protection strategy'. Since a quarter of all energy use is in residential properties, this is a logical stance and ought to be shared by other countries. The intention behind the policy is to improve the energy efficiency of the housing stock by 3 per cent a year. It is arithmetically obvious that this cannot come from new buildings, which each year form at most 1 per cent of the total number of houses. The bulk of the improvement will therefore come from the refurbishment of the oldest houses, particularly those built before energy efficiency rules on new buildings were imposed in the late 1970s.

The primary mechanism used by the Federal Government is a subsidised long-term loan scheme for any owner wishing to reduce the energy use of the building. The Germans say that pre-1984 homes have three times the energy use of buildings constructed to today's energy efficiency regulations. Typically a refurbishment aimed at reducing energy use only achieves half of the savings that can be reached. But a really effective eco-refurbishment can reduce energy use by over 80 per cent, taking the energy consumption of many older homes well below the standards currently demanded for new buildings. It is a mistake to assume that only building new homes can give us good insulation performance.

The German Energy Agency started a project in 2003 to refurbish 140 buildings to prove this point to a sceptical construction industry. Many of these buildings contain multiple apartments and in total the study looked at over 2,000 separate homes. The Agency claims that the refurbishments have cut energy use so much that these buildings now have electricity and heat consumption of less than half what would be expected in a conventional new-build.

The Passivhaus Institute agrees about the potential for refurbishments. It says that it is possible to achieve energy efficiency results in an eco-refit that are not far behind the achievements of the best passivhaus new homes. The Institute quotes a figure for the energy used in heating of about 25 kilowatt-hours per square metre for the best refurbishments, compared with 15 for a passivhaus. That may not seem particularly impressive until you realise that for a typical thirty-year-old European house, the figure is over 200. (Flats and apartments are lower because of the insulation effect provided by shared walls.) For an average thirty-year-old home, getting down to 25 kilowatt-hours per square metre would mean a saving of about 3 tonnes of carbon dioxide each year.

One astonishingly successful renovation in Linz, Austria, should convince us all that the potential savings from converting old apartment blocks are a much better way of spending money than just insisting on ever-higher standards in new building. This fifty-unit apartment block was built in 1957 and had heating energy use of about 180 kilowatt-hours per square metre. The refurbishment a few years ago reduced this figure to below 15 kilowatt-hours. The other advantages included increasing the internal size of the properties by over 10 per cent and significantly reducing the traffic noise inside the apartments. Air quality in the homes is better, as incoming air, polluted by vehicle fumes, is now filtered. The cost of this path-breaking eco-renovation was nearly €50,000 per apartment, so the

Apartment block in Linz, Austria, before and after renovation

reductions in energy use are only likely to pay back the investment over several decades, but the improvement in the appearance and livability of the block is striking.

If improvements of this size were delivered across the whole housing stock, the prospective savings would be at least 10 per cent of national emissions and possibly much more. A very welcome by-product would be a reduction in the number of people suffering from cold and discomfort in winter. Though the number is tending to fall, partly as a result of climate change, more people die in the winter in northern Europe than in the summer – around 20,000 more in the UK, despite the country's relatively mild winters. Some of the reason for this difference is that older people are more likely to get ill in inadequately heated homes.

A variety of approaches were used in the 140 buildings of the crucial German study, with some of the refurbishments making extensive use of highly innovative passivhaus components for insulation and air-tightness. The results have helped shape the subsidised loan scheme now available to the general public. The German national and state governments offer a substantial incentive to achieving really good energy performance. If a renovation achieves energy use 30 per cent below the building regulations for new-builds, the government will forgive 12.5 per cent of the money owed. Smaller energy savings are rewarded with smaller, but still significant, savings.

All told, the German government is ploughing over €5bn per year into the eco-refurbishment of existing houses in the form of loans and subsidies. This seems a large amount, but one estimate puts the total amount of money spent on all types of improvement of the structure of houses at nearly €80bn in Germany alone. The eco-loan scheme has been popular with landlords wanting to maintain the standards of their stock of buildings, particular those operating in the social housing sector. In 2007 the scheme helped reduce energy use in about 200,000 homes, or about 0.5 per cent of all the houses and flats in the country. This number is almost as great as the total number of new housing units being built each year. So, unlike Britain and the other countries that are focused on regulating the efficiency standards of new buildings, Germany is seeing far greater carbon dioxide cuts from putting money into the oldest portion of the housing stock. The sponsoring government ministry says that each year of the programme has saved annual emissions of about 1 million tonnes of carbon dioxide, or about 5 tonnes per housing unit renovated.

Does the German scheme make financial sense, both to the homeowner and to the government? On the basis of the numbers released by the government, the answer is a cautious yes. If we assume the refurbishment lasts fifty years, then spending a billion euros to cut emissions by 1 million tonnes a year implies a cost of €20 for each tonne saved, a highly competitive figure. And since this money is not a direct subsidy, but simply a loan, the effective cost is far, far smaller.

The impact on the householder is less clear-cut. The savings in fuel bills arising from the refurbishments are probably between €500 and €1,000 a year, and often less. An investment of up to €50,000 to achieve these cost reductions is not an obviously successful investment with an annual saving return of perhaps only 2 per cent. (The central government is more optimistic, quoting typical returns of several times this level, but their figures seem unusually high.) Nevertheless, what is absolutely clear is that landlords and homeowners are extremely keen to carry out energy efficiency refurbishments, whatever the short-term benefit in reducing gas and electricity bills. Refurbishment makes the home more comfortable and, for owner-occupiers, much easier to sell.

A secondary benefit to the German economy has been an upsurge in the number of jobs in the construction industry. Germany has suffered from high unemployment, particularly in the states of the former East Germany, so this side-effect has been extremely welcome. The government quotes figures of more than 200,000 people pulled into permanent employment as a result of its refurbishment programme. A US study by the Center for American Progress in September 2008 suggested that a strong green stimulus might replace all of the 800,000 jobs lost in construction in the previous two years, or about the same percentage of the working population as in Germany.

About three-quarters of German housing was built before the end of the 1970s, when insulation standards began to improve. To refurbish all these houses within the next thirty years would mean tripling the current rate of refurbishment. That sounds ambitious, but the eventual savings may approach 100 million tonnes a year of carbon dioxide – or about 10 per cent of the German total.

The technologies for improving the energy consumption of domestic housing are simple and relatively well understood, partly as a result of the pathbreaking work of Wolfgang Feist and the Passivhaus Institute. Alongside more exciting and more glamorous techniques for carbon reduction, the world needs to devote efforts to making the easy gains in house insulation and airtightness. So far, developed countries have been slow to see the potential and cost-effectiveness of refurbishment. But as energy costs rise and more and more families struggle to meet heating bills, the obvious scope to reduce energy use in homes provides an attractive target for massive emissions reductions. And we could start tomorrow.

Electric cars

The inevitable switch to battery propulsion

In October 2007, the Israeli-born Californian software engineer Shai Agassi announced a plan to accelerate the move to the all-electric car. Agassi is a visionary with a capacity to inspire change. But the scale of the task he took on is breathtaking. At the moment, car batteries for electric-only vehicles are extremely heavy, recharge slowly, have a limited range, and are eye-wateringly expensive. Agassi's scheme requires car makers to build a new generation of all-electric cars, probably with no back up engine. His company will then offer to rent batteries to the car owners at a cost that beats the price of petrol. Owners will recharge the cars at home or at work. For long journeys, they will simply swap the batteries at automated recharging stations conveniently situated on major routes.

Agassi has taken on one of the most important challenges facing us: the need to create a world vehicle fleet powered by an alternative to globe-warming petroleum fuels and inefficient internal combustion engines. We can't know whether his venture will succeed. But he is just one of the many people betting that petrol and diesel will be replaced by electricity. Battery technology is likely to advance very substantially in the next decade. Price, weight, range and recharging time will all improve. In contrast, although the price of oil has subsided since the mid 2008 peak of $140 a barrel, oil is likely to get more expensive as time passes. Electricity is now very clearly a less expensive way to run a vehicle and is likely to become even more attractive in the future.

Small numbers of all-electric cars have been around for decades. They are generally slow and have a very limited range. Perhaps

Shai Agassi

25,000 of these vehicles putter around the suburban streets and retirement communities of California and other places in the US. But battery vehicles that look and drive like ordinary cars will be on sale in late 2010 with mass production from manufacturers such as Chevrolet and Nissan slated for 2012. Batteries will need to be recharged with electricity, and at the moment that electricity is most definitely not 'zero-carbon'. But many of the other advances suggested in this book will substantially reduce the carbon produced when electricity is generated. As time progresses, the advantages of using electricity to drive our cars can only grow.

The modern car is the product of more than a hundred years of evolution. It is extremely reliable, at least compared to its ancestors. Usually comfortable and pleasant to use, it is also much safer than it was even twenty years ago. But a century of improvements hides a surprising fact. The engines that power today's cars remain inefficient and wasteful. Automobile manufacturers spend enormous sums each year on research and development, but only about a quarter of the energy in petrol actually gets to the wheels of the car. Over three-quarters is wasted as heat, mostly from the exhaust or through the cooling system. Diesel cars are better, but still waste most of the energy in the fuel.

We will see some small improvements in internal combustion engines over the next decades. The cars hurtling round the track at a Formula 1 Grand Prix turn about a third of their fuel's energy into useful motion. The billions spent every year trying to take milliseconds off lap times will eventually result in technical advances that can be used in vehicles whose most exciting ride may be a trip to the supermarket. But internal combustion engines are never going to be parsimonious in the use of fuel. By contrast, electric motors can turn 80 per cent of the power delivered by a battery into useful motion. This simple fact means that it is not a question of *whether* the electric car will take over the roads, but rather a question of when.

The answer to this question has important implications for climate change. As more and more people around the world become rich enough to afford a car, the emissions from automobile engines increase. There are about 600 million cars on the road today. The burgeoning Chinese and Indian middle classes might push this number up by several hundred million in the next decade alone. Cars and lorries are already responsible for about 15 per cent of global emissions. Without a revolution in technology, this number will inevitably rise, even if car manufacturers surprise themselves with the improvements that they can eke out of Nikolaus Otto's 1876 designs for a four-stroke compression engine.

Of course, cars only represent part of the problem. Our roads are also used by vans and lorries that transport our food to the supermarkets, our clothes to shopping centres, and our waste to landfill sites. It is more difficult to convert these vehicles to batteries. Their power needs are too great and many of them make longer distance trips that would run down even the largest battery pack. Some commercial vehicles can be converted to fuel cells: city buses are probably the best example. Other commercial vehicles can run on second-generation ethanol as a substitute for petrol (see Chapter 7). But large lorries making long journeys will need to use liquid fuels until we find a way of making diesel from agricultural wastes. Chapters 9 and 10 look at the ways in which we can extract carbon dioxide from the air to offset the emissions from those vehicles that we can't easily convert to electricity.

Why electric?

To some people, the answer to slashing transport emissions is simple. We should engineer our world so that public transport replaces private cars. We should strive to reduce the distances to our workplaces and our shops so that people need to drive less. If we 'localised' our food production we might be able to omit our weekly drive to the out-of-town supermarket and cut the 25 per cent of all lorry journeys that, in many countries, are accounted for by carrying food to shops, factories and warehouses. On those occasions that people still need to use a car, they should have access to pools of shared small cars.

These policies are all sensible. But the unfortunate fact is that the private car is a possession of high value to many people: it provides freedom, independence and, all too often, status. Measures to damp down car ownership are fighting against strong human desires. Yes, we can reorganise our cities to make bus travel easier and more comfortable, and we can make cycling safer and more fun. But such measures will make only a small dent in the demand for petrol and diesel.

To make a bigger difference, we clearly need to produce the greenest possible cars. There are various options on the table for achieving this: the use of hydrogen as a fuel; smaller and more efficient conventional engines; low-carbon fuels from agricultural products ('biofuels'); and electric cars.

Despite the sympathetic hearing sometimes given to the first of these alternatives, we can quickly dismiss it. There are three enormous problems with hydrogen. First, it would require a completely new set of refuelling points. Thousands of filling stations would need to be established in every country. The cost of hydrogen storage is high, not because it is dangerously explosive (in most circumstances it is not) but because it needs to be stored at high pressure. At low pressure, hydrogen occupies a substantial amount of space for each unit of usable energy it contains. So it needs to be hugely compressed so that the storage tanks do not take up too much space. High-pressure storage is expensive and the cost of building the tanks necessary to hold the gas would be prohibitive.

The second problem with hydrogen is that it takes substantial amounts of energy to make it in the first place. The easiest way of creating pure hydrogen gas is to split water into its constituent atoms, oxygen and hydrogen. This is a simple process, but one which takes substantially more energy, usually in the form of electricity, than can be usefully captured when the hydrogen is later burnt in a car engine. This lack of any energy benefit is an unchangeable law of chemistry, and nothing is going to get round the problem. Of course, the electrical energy we use to make the hydrogen could come from renewable sources, in which case there would be no carbon dioxide cost. But we could have used this electricity for other purposes instead. Making hydrogen with wind power means that we cannot use the electricity to power homes and offices, or indeed to refuel car batteries. A kilowatt-hour of electricity in a battery will drive a car far further than the same amount of electricity used to make a tank of hydrogen.

Third, cars that use hydrogen as their power source only exist in very limited numbers. The Honda FCX (Fuel Cell eXperimental) is on sale in some parts of southern California that are within easy reach of the small number of hydrogen refuelling stations. A few other prototype hydrogen cars are also on the roads. There are two different approaches that manufacturers have taken: the hydrogen can be burnt to capture energy, much as in a conventional engine. Or the hydrogen can be used to generate electricity in a fuel cell inside the car – this is the technology used in the Honda FCX. As Chapter 4 explains, fuel cells are a very promising technology for static heat and power generation, but they're currently expensive and too large for use in cars. One estimate is that each of the Honda fuel cell cars cost almost a million dollars to make. They are also not obviously suited to the vibration and rough treatment meted out to an engine as it travels over uneven roads. Fuel cell vehicles exist, but the technology is probably more suited to sedate urban buses than cars that need travel a long way from their home refuelling point.

The second possible route forward is to make lower-emission cars. Governments around the world are focusing on encouraging manufacturers to build smaller, lighter and more efficient cars. They are applying pressure both by setting maximum limits on the typical emissions of each car company, averaged across all their models, and by increasing taxation on the vehicles that consume most petrol, either through fuel taxes paid on each litre, or by increasing excise duties on bigger cars. Proposals from the European Union to mandate further improvements in emissions standards have elicited howls of pain from the manufacturers of big and heavy cars, such as Mercedes-Benz, which will have to make disproportionately large reductions. In late 2007, the US congress similarly approved emissions targets for automobile manufacturers, which will force them to make substantial improvements in fuel economy. This is only after thirty years of gradually decreasing fuel economy in US cars.

The problem is that even the smallest, lightest, most aerodynamic petrol-powered European or Japanese cars emit around 100 grams of carbon dioxide per kilometre travelled. Given that the average car in Britain travels about 9,000 miles annually, even the best new car will typically emit about a tonne and a half of carbon dioxide a year. Cars in the rest of Europe travel shorter distances and use slightly fewer litres of petrol, so that carbon dioxide emissions are lower. But the figures are far higher in the US or Canada. Given that the world will probably need to cut total emissions across all activities to no more than one or two tonnes per person by 2050, cars powered by internal combustion engines represent an important target for carbon reduction.

Governments have also been keen on biofuels, the third option for cutting car emissions. Carbohydrate crops, such as wheat, maize or sugar beet, are distilled into ethanol and oil-bearing products, such as oilseed rape seeds or the berries of the tropical plant jatropha, are turned into diesel. As the following chapter explains, however, official enthusiasm for vehicle fuels from crops is being quickly eroded by the increasing evidence that fuels made from food crops do little to cut emissions.

The next chapter looks in detail at second-generation biofuels made from agricultural and forestry wastes. The scope for using these fuels for reducing emissions is much greater than with today's crop-based ethanol and there is considerable reason for optimism about the long-run potential of low-carbon liquid fuels. But even the new cellulosic ethanol fuels need truly enormous amounts of feedstock from forest and fields. We may never be able to fuel our needs for personal transport without causing further problems of deforestation and the loss of food-producing land. It makes good sense for policymakers to encourage car owners to use electric vehicles. Although the move to a car fleet that is very largely powered by electric batteries has many challenges, it will eventually reduce driving costs and substantially cut carbon dioxide emissions.

Some automobiles that can switch between electric batteries and conventional petrol engines are already on the road. These cars are called 'hybrids' and the best known is the Toyota Prius, a good-looking medium-sized saloon car. The Prius's batteries power the vehicle around cities. On longer journeys the petrol engine kicks in and the car operates as a conventional motor vehicle. The car's batteries are recharged by recapturing the energy lost from the vehicle as it brakes. Think of it this way: a car travelling at 30 mph has a lot of kinetic energy. As it slows to a stop at a traffic light, this kinetic energy is lost. The law of conservation of energy says that the power of the moving car must be translated into heat, or some other equivalent energy carrier. Most vehicles turn deceleration into heat – this is why brakes get hot. A Prius is different. It captures some of the energy from braking and turns it into electrical energy in a battery, a process known as 'regenerative braking'. The same principle is now also used in some trains and a small number of other cars.

The Prius's combination of internal combustion engine and battery has several other advantages. First, the battery can be used to move the car forward after it has stopped at junctions or traffic lights. This means that the petrol engine can be turned off when the car is stationary, rather than wasting fuel when idling. Second, the battery can work in tandem with the engine to increase the power of the car when it is accelerating. This means that the combustion engine does not have to be as large as it would normally be in a car of this type. Since small petrol engines deliver inherently better fuel economy than larger ones, this reduces the average petrol consumption of the car.

As a result of having the battery available, the Prius's fuel economy was outstandingly good when it came out five or so years ago. Its carbon emissions were as low as any standard car on the road. But other manufacturers have made their own improvements, and several small cars now have fuel consumption that nearly matches the Prius. For example, the Toyota iQ, a small car sold in Europe, has carbon emissions of 99 grams per kilometre, only slightly more than the 89 grams of the Prius. The iQ doesn't do anything sophisticated like capture kinetic energy and turn it into electricity. It is simply a light car that needs limited power and its small engine is good at turning the chemical energy in fossil fuel into movement.

The Toyota Prius and other hybrid cars are the first steps on the road to fully electric vehicles. They have batteries that can only carry the vehicle for a short distance and their electrical propulsion system has to operate alongside the conventional internal combustion engine. This makes the car expensive, complex and heavy. The battery of the Prius weighs about 50 kilograms and stores less than 1 kilowatt-hour of energy, approximately enough to drive the car at a constant speed for about six miles. Put another way, the battery is only able to accelerate the car from standstill to 60 mph four times before it needs recharging.

Within a few years of Toyota bringing out the Prius, 'hackers' had spotted an obvious improvement to the car. They added an informal modification that allowed the owner to plug the car into the electricity mains to recharge the battery from the grid rather than through regenerative braking. These people have turned their Prius cars from being ordinary hybrids to being 'plug-in hybrid electric vehicles', or PHEVs in the specialist jargon. The owners can plug the car in at night and use the batteries to drive short distances to their job every day. On longer trips, they can still use the petrol engine to power the car when the batteries run down.

Many people think that PHEVs are the car of the future because they combine the advantages of battery cars and conventional engines. A very large fraction of all car journeys cover only short distances, so for many people a car such as this would almost exclusively be powered by electricity. The internal combustion engine will kick in during the rare longer trips. The owner will have the certainty of knowing that the car will have a source of power even on an unexpected trip across the country.

But PHEVs also have disadvantages. The car's batteries are heavy and expensive. They have to fit in the car alongside the petrol engine and its transmission system. Putting both an internal combustion engine and an electric propulsion into the car is very costly and reduces storage space. The pioneering Massachusetts battery company A123 Systems is selling an additional battery pack for the Prius that makes it far closer to an all-electric car. At $10,000, it's an expensive add-on, but for regular users, commuting every day for fifty miles, the battery may just about make financial sense.

A battery-only car could, in theory, be a much better idea than a hybrid like the Prius. Although the batteries would have to be even larger than in a PHEV, the cost and space savings from not having an internal combustion engine could be very substantial. An all-electric car might simply have electric motors driving each wheel and almost all of the complex system of gearboxes, cooling systems and power transmission would disappear. As battery technology improves, it will eventually become much cheaper to build an electric car than even the simplest fossil fuel alternative. The only surprising thing is that this hasn't happened already.

The other advantage, of course, is that electricity is much cheaper than petrol, particularly in countries with punitive fuel taxation, such as most European states. The electric car is able to convert a large fraction of all of the energy in a battery into motion, compared to about 20–25 per cent for a petrol equivalent. A light electric car travelling at 40 mph uses about 7 kilowatts of power. At current UK electricity prices, the cost of this is about 80 pence an hour. Even a highly fuel-efficient small petrol car will cost three or four times that amount at current UK petrol prices of over £1.20 a litre.

About half the cost of petrol in the UK is taxation, and the advantages of electric cars in the US and other low-taxation countries will be smaller. But in almost all countries an electric power source will cut motoring costs by over two-thirds. The relative simplicity of all-electric cars will eventually also mean reduced maintenance and insurance costs.

So from the car owner's point of view, electricity is better than fossil fuels such as diesel and petrol. And although electricity may get more expensive because of rising fossil fuels prices and the expensive subsidies given to renewable electricity generators, it is likely to remain considerably cheaper than liquid hydrocarbon fuels, even when full-scale production of next-generation biofuels begins in a few years.

What about the impact of electric cars on carbon emissions?

Electricity from the grid is most definitely not carbon-free. In the UK, running a 7-kilowatt electric car for an hour will produce about 3.5 kilograms of carbon dioxide from power stations when the battery is recharged. Other countries will be higher or lower, depending on how they produce their electricity. In a country like the US, which generates most of its electricity from coal, an electric car powered from the mains would produce more carbon dioxide than in France, which gets most of its power from nuclear stations, or Denmark, which has wind energy and imports hydroelectricity from other Nordic countries.

An efficient petrol-powered car, by comparison, would probably produce about 10 kilograms of carbon dioxide if driven at 40 mph for an hour. A hybrid wouldn't be much better. So an electric car has emissions of less than a third of a new petrol car on the road in countries using fossil fuel to generate most of their electricity. As industrial countries gradually increase the percentage of electricity coming from renewable sources that do not produce carbon dioxide, the advantages of powering our transport fleets with batteries will become even more obvious. The financial and ecological arguments for using electricity to power our cars are overwhelming. And there are other advantages. Electric cars produce no pollutants when driven around, so the air quality in cities would be much better, helping reduce the incidence of asthma and respiratory diseases. Most electric vehicles are also very quiet, so a large-scale switch would reduce the noise level around busy roads. The evidence that high noise levels affect human health is not yet as convincing as the link between city centre traffic and asthma. Nevertheless, the standard of living of millions of people would be improved if we reduced the background level of traffic noise in industrial countries.

The roadblocks

I suspect that eventually almost all our cars will be electric. But getting from a position in which almost all vehicles are powered by grossly inefficient internal combustion engines to one in which lighter cars glide noiselessly around our cities is not going to be easy.

We face four main problems. First, batteries are expensive. Today, a power pack that will drive a car for a hundred miles will cost over $12,000. This might double the construction cost of a small family car. Second, the batteries weigh 200 kilograms or more, adding 20 per cent or more to the weight of a small vehicle. The time needed to recharge a battery is declining, but it can still take eight or ten hours if charged at home. And, lastly, batteries for cars will probably be based on a light metal called lithium which is only mined in a small number of places around the world, mostly in South America. Lithium is the key element in today's batteries for mobile phones and laptop computers. Although there is no current shortage of the metal, a substantial shift to electric cars would increase worldwide demand by a multiple of a hundred. If we need to make tens of millions of car batteries every year, there will be inevitable problems getting reliable supplies at a reasonable price.

There are other minor obstacles. One problem is that it is surprisingly difficult to tell what percentage of a battery's charge remains. Users will be understandably cautious about adopting the electric car if they do not know how far they can drive without a top-up of electricity. In the past, batteries would typically fail after a relatively small number of charges. Even today, lithium-based laptop batteries often cease to work some time before the rest of the computer fails. But recent improvements should mean that new batteries will last at least as long as the cars they're found in, even if they are charged and completely discharged every day.

There's another potential obstacle that rarely gets noted. Governments in Europe and elsewhere get a large fraction of their tax revenue from fuel duties and vehicle taxes. In the UK, for example, about 6 per cent of all government receipts are generated this way. Will governments have the courage to lose such a valuable source of revenue, or will they quietly discourage electric car use? The problem is likely to bite first in London, where the city's revenue is boosted by the tax charged on vehicles entering the central area. At the moment, electric cars go free and, in some places in the city, can even be recharged at no cost to the driver. Will the inevitable rise of the electric car be held back by London mayors cautious about the impact on their revenue? As the number of electric cars grows, will they be able to avoid the temptation to start taxing them?

The Tesla electric sports car

Past history does not inspire confidence that national and local governments will encourage technology – however green – that cuts sharply into their tax base. But because electricity is so much cheaper than petrol, there will probably always be an incentive to switch to a battery-powered car.

What about acceleration and top speed? The conventional supposition – that electric cars are necessarily slow and sluggish – is wrong. New types of battery can deliver explosive amounts of power. The much discussed Tesla Roadster, an electric sports car, finally went on public sale late in 2008. The car accelerates from standstill to 60 mph in under 4 seconds, with no noise and no gearbox strain. The Tesla is also very fast: the roadster has a top speed that has been electronically limited to 125 mph. Designed and partly assembled by Lotus Cars in the UK, the Tesla is an astonishing piece of automobile engineering that has helped change the image of electric cars around the world. The price tag is over $100,000, but drivers will get a car with a performance that matches the best petrol-powered competitors. Incidentally, it is worth mentioning that the car has a range between recharges of over 240 miles, largely because the manufacturer has opted to install batteries of very high capacity. This is one of the prime reasons why the car is so expensive.

At $100,000, the Tesla is not a car for the ordinary family. How will the large automobile manufacturers around the world get to the point where electric cars become the standard offering in their showrooms?

The route to the electric family car

California is the centre of the electric car industry. But even there, buyers face a limited choice. Those who cannot afford the Tesla, or who are a little too sedate to fully use its extraordinary acceleration, are left with vehicles that barely reach 25 mph and are only really usable for dawdling to the local shops on quiet side roads. Their owners are happy to live with their limitations, but these electric runabouts are not real substitutes for the flexible petrol saloon. General Motors tried to introduce an electric car in California in the late 1990s, but its attempt failed, largely because of the high cost of the car and the limited range of its first-generation batteries. Many people still say that the demise of this car, the EV1, was hastened by the opposition of the oil industry, but the unfortunate reality was that this vehicle was simply not good enough – it didn't have the performance of the Tesla or the simplicity of the glorified electric golf carts that trundle around some California communities.

The second major market for electric cars is in England. Fuel taxation is high by international standards, so electric propulsion looks particularly good value. Car owners also have to pay a yearly tax on their vehicle but low-carbon cars pay little or nothing. Perhaps most importantly, the Congestion Charge imposed on vehicles entering and leaving central London exempts electric cars. Unsurprisingly, then, the world's small band of electric car manufacturers has made the city a focus for their sales efforts. The Indian manufacturer of the G-Wiz has sold over 1,000 of its spectacularly ugly electric cars to London's commuters. Until recently, this manufacturer used the old-fashioned lead–acid battery (the type that powers the starter motors in petrol cars) and this limits the range of the car to a few tens of miles. Top speed is theoretically more than 40 mph, though it can be much lower when going up London's few hills.

Such limitations may not matter much in central London, where driving distances are short and, as the company selling the G-Wiz points out, travel speeds barely exceed walking pace, even if you are in a Tesla. But for the average driver outside congested city centres, speed, acceleration and range do matter. Thankfully, the technology is improving fast and it should soon be possible to power a reasonably sized car for a long commuter journey with batteries that aren't prohibitively expensive and which can comfortably accelerate the vehicle to highway speeds. Early lead–acid batteries had neither the power nor the storage capacity to do this, and neither did the second-generation 'nickel metal hydride' technology, as found in the Toyota Prius. Only with new 'lithium-ion' batteries has good range and power been possible.

Unfortunately, current lithium-ion technology has two significant problems – expense and a propensity to explode if improperly manufactured or mechanically abused. The chemistry in lithium-ion cells means that the stored energy can rapidly be released in the form of heat under certain circumstances. In 2006, manufacturing flaws that left some impurities in batteries for laptops meant that a small number caught fire and many tens of thousands were recalled by the manufacturers. When fully charged, lithium-ion batteries in cars might have a thousand times more energy inside them than is stored in a laptop. So safety measures have to be effective and reliable.

The current consensus is that this small risk of explosion may mean that first-generation lithium-ion cells are not the most appropriate technology for mass-market cars. An alternative, usually known as 'lithium iron phosphate' (LiFePO4), may well be the future choice but a variety of similar technologies are fighting for market acceptance. Compared to their lithium-ion predecessors, LiFePo4 batteries are slightly heavier for each unit of power that they contain. But they will not explode or catch fire, they can be charged quickly, and they will work for thousands of charging and discharging cycles. LiFePO4 will eventually be cheaper than conventional lithium-ion, too, partly because the new technology uses iron, an inexpensive metal, where lithium-ion batteries require expensive cobalt.

Nanotechnology, the science of materials at extremely small scales, should enable future batteries to charge more quickly. By changing the structure of the anode of the battery, almost at the level of individual molecules, we are likely eventually to be able to recharge a car battery in minutes rather than hours. This improvement has not yet got far beyond the university laboratory, and it remains to be seen whether the technology will ever be cheap enough for the mass-market, but it's possible that within five years we will be recharging car batteries almost as fast as we fill up a petrol tank today.

Unfortunately, technological advances will not necessarily improve the storage capacity of batteries. There are real and unbreakable rules about how much energy can be held in each kilogram of each type of battery. We know that if $LiFePO4$ continues to be the technology of choice, then we will always need approximately 6 kilograms for every kilowatt-hour of electric power. A kilowatt-hour will take a standard small car about four to five miles, depending on its weight and the amount of acceleration used. So for every mile of range we will need more than a kilogram of battery. There's no reason to suppose that we will not develop lighter batteries in the future, but there are no obvious breakthroughs on the horizon.

Automobile manufacturers and battery makers face a difficult task in finding the right balance. Do they put heavy batteries in an electric car, increasing its range? Or do they reduce the weight, keeping the battery cost down and adding to the space for passengers and luggage, thereby limiting the distance the car can travel between recharges?

Perhaps the answer is that the car companies will offer two types of vehicle. Some models will be sold with smaller batteries that enable the car to be driven to work, recharged there, and then driven home. After all, the average British car is only driven 25 miles a day and the typical American commuter journey is about 25 minutes. So for most car owners a battery that lasts, say, 40 miles is going to be perfectly adequate for most journeys. If they need to travel further, they can borrow a car with a greater range from the company car pool or a local car-sharing club.

The optimists in the industry believe that the cost of batteries for a car of this type will fall to about $150 per kilowatt-hour in the next decade, or somewhat less than $1,500 for a range of 50 miles. This is at least a four- or five-fold reduction. There will also be a cost for the electronics needed to control the battery's charging cycles. The manufacturer will save by not having to include an engine and a transmission system in the car, but an electric car is likely to cost somewhat more than a petrol equivalent for some time. But since the yearly running costs will be a half or a third of existing levels, this isn't necessarily a major obstacle.

Users who frequently drive longer distances will need a car with more battery capacity, or possibly will decide to buy a hybrid vehicle. General Motors (GM) has decided that the way forward is to add a small internal combustion engine to its Volt electric car, due to be launched in late 2010. When the battery is running down, the vehicle's 1.0-litre petrol engine will kick in and begin to provide electric power to the motor. The engine will never directly drive the car, unlike with the Toyota Prius, which switches between electric and petrol propulsion. GM's solution – giving the engine the simple role of on-board electricity generation – seems simpler and more elegant than the traditional hybrid design, and should help minimise production costs. The battery pack in this vehicle is quite small – only 16 kWh – and the manufacturer says that it will only be allowed to run down to 30 per cent charge. This means its range (without topping up with petrol) will only be about 40 miles, or enough for about 70 per cent of US daily commuting journeys. The price of the car is higher than petrol-driven equivalents, but the cost is reduced by a $7,500 government subsidy.

Nissan is also intending to launch an electric car in 2010. Its LEAF model will aim for 100 miles between charges, with the battery containing about 24 kWh. The company has talked about leasing the battery to users for a fee of about $150 a month. It seems a lot, but the net cost will be lower than most people's fuel savings. Nissan's 2010 cars will be manufactured in Japan, but as full production ramps up, its Tennessee factory will start also assembling the vehicle.

And what about the scheme proposed by Shai Agassi in the first paragraphs of this chapter? Agassi's insight is that users will be put off by the higher initial costs of electric cars but will be willing to pay to rent the batteries out of the fuel savings that they make. His scheme requires car manufacturers to design vehicles that offer

Think city cars

easy swapping of discharged batteries with freshly charged replacements. Renault-Nissan has committed to work with his company. He also needs to persuade local entrepreneurs or governments to install a network of swapping points that give long-distance drivers the security of knowing that they can get new batteries when they need them. This is an enormous challenge and, given its huge size, the US may not be the best place to start. Instead, Agassi seems to have persuaded the Israeli and Danish governments to back his scheme. The smaller size and shorter travel distances of these countries make the goal more achievable. Renault Nissan has promised to make at least 100,000 electric cars available by 2016 in these two countries, thereby providing a market for the new charging stations and battery swapping points. Israel's central position in Middle Eastern conflicts makes the country unusually vulnerable to loss of its oil supply, so the government has a strong incentive to support Agassi's vision over the long term.

Agassi's business has also signed up supporters in Denmark. His partner there is Dong Energy, which runs many of the offshore wind farms in the country. Dong will supply renewable energy for the batteries, meaning that the owners of the cars will know that their driving is genuinely low carbon. On those occasions when Dong is generating too much energy to be used on the Danish grid, it will also make obvious financial sense to use the spare capacity to charge car batteries. Similarly, Agassi's Israeli partners are intending to use solar energy from the Negev Desert.

Meanwhile, the Norwegian company Think has launched its all-electric city car in Europe. This vehicle has a range of over 120 miles and a top speed of 60 mph. The batteries can be completely recharged overnight. Eager to show that this is a 'real' car, the company has constructed the vehicle with a steel cage to meet international safety standards. This is not an electric golf cart. Even though its acceleration leaves a lot to be desired, this is a car that can compete against petrol equivalents. Early reviews in the UK were a little patronising, focusing on some of the less-sophisticated features of the vehicle, but Think is delivering a product very close to the performance of a conventional compact car. At about £14,000 in the UK, the price is forbiddingly high and costs will need to come down sharply if it is to achieve large numbers of sales. But since the Think car qualifies for exemption from the £8 per day London driving tax and driving into London every working day costs over £2,000 a year, some people will be prepared to pay the hefty price for the car.

Cars like the Think demonstrate the opportunities and threats to Agassi's and Renault-Nissan's plan. On the one hand, by separating the ownership of the car and the car's batteries, Agassi's business could reduce the purchase price of an electric car substantially. But if my earlier prediction is correct and we eventually see battery recharging times falling to five or ten minutes then Agassi's scheme will fail. It won't make sense to design cars to have easily removable batteries and to install heavy machinery at hundreds of locations to take them out and automatically replace them. If charging can be completed in five minutes we will simply get used to driving into a 'filling' station, plugging the car in, and having a coffee. And if electric cars remain too expensive because of the up-front cost of the battery, then car-leasing companies will come forward to offer better financing terms. The world may not need Shai Agassi's scheme for kick-starting the electric revolution after all.

Cars and the grid

There will be substantial obstacles on the way to battery-based driving, but the economic and environmental arguments are too compelling for the electric car not to eventually win the day. Electric cars will save drivers money and, with private cars and light commercial vehicles responsible for 20 per cent of European carbon dioxide emissions and more in the US, they represent a big step forward on the road to a low-carbon future.

Even if vehicles are charged by electricity made at a coal-burning power plant without carbon capture, they will save emissions. But that, of course, is not the vision. We want cars to be charged by electricity from renewable sources, and to work symbiotically with renewable power sources. The proposed joint venture between Shai Agassi and Dong Energy in Denmark is a good example of this. In the longer term, we want car batteries to act as electricity storage for the grid. Renewable electricity can be intermittent (as with tidal energy) or both intermittent and unpredictable (as with wind). As described in Chapter 1, car batteries linked to the grid with intelligent communications could stop charging when supply falls, and even feed energy back into the grid when necessary.

A million electric cars parked at homes or offices, and plugged into the mains, could meet 5 per cent or more of the electricity supply in a big country such as Germany, possibly within a few seconds. This might double the emergency buffer of electric power held by the operators of national electricity grids. The value of this would be enormous, both in allowing the country to use more intermittent and unreliable energy sources and in reducing the need for fossil fuel power plants to be kept on standby. It might also significantly improve the attractiveness of owning an electric car, because utilities will be prepared to pay for the value of the storage capacity in the battery. (Rather optimistic estimates produced by Hillary Clinton during the 2008 US primary campaign even suggested that the financial value to the utilities of having access to the batteries would be greater than the cost of the batteries themselves. If this were true, then it would make sense today for the electricity companies to be buying their own batteries!)

Turning car batteries into emergency stores of power is all well and good, but electric cars are still ultimately consumers of electricity. So will a switch to battery-powered vehicles put an impossible extra demand on power generators? Shai Agassi says that if every car in Israel were powered by electricity, it would only add 6 per cent to national electricity demand. Rough calculations suggest that the figure might be about 12 or 15 per cent in the UK. These numbers suggest that the extra electricity use can easily be met from renewable sources such as solar power in Israel, wind in Denmark or tidal power in the UK. With intelligent charging systems, we will only be sending electricity to the batteries when there is a surplus in the national electricity distribution system. Recharging will be mostly carried out at night when demand is relatively low and most cars are parked next to domestic homes with easily available power sockets. When power is short, charging will cease and the flow of electricity will be reversed as batteries help to balance electricity supply and demand. Because we can be sure that charging car batteries will not add to the daily peak demand for power, we will not need to construct new power stations. However, we do need to be aware that there may be some times when it becomes impossible, or extremely expensive, to charge a car's batteries because the grid is unable to produce enough power. At these moments, Shai Agassi's battery rental scheme will come into its own. The car owner would drive to the local battery exchange point and swap his or her partly charged units for full equivalents.

Of all the technologies in this book, battery driven cars have advanced furthest in the last year. The flow of news during 2009 has been remarkable, with many major manufacturers announcing the development of new ranges of electric cars.

By late 2010 or 2011, most vehicle manufacturers will offer a range of plug-in hybrids or even electric-only cars. Optimism should be tempered by the fact that hybrid cars are still only 3 per cent of US sales in autumn 2009, perhaps because of the large price premium, but electric cars are going to make rapid advances, buoyed by rapidly declining battery costs and the prospects of higher petrol prices in the future.

The twenty or so major car manufacturers are increasingly aware that for the first time ever their dominance of world markets faces a powerful threat. The internal combustion engine and complex mechanical transmission and braking systems of a typical car are eventually going to be considerably more expensive than simple and reliable electric cars. Despite their huge manufacturing scale, marketing skills and worldwide network of retailers, these companies could be rapidly undermined by new and fast-moving competitors such as Tesla and Think. Growing awareness of this fact has pushed the global car manufacturing industry into a gear-change. After dismissing the prospects for battery cars for decades, the car manufacturers are beginning to invest in their own electric vehicle projects. This is excellent news: we need these companies, with their huge resources and skills, to get behind the electric car. As with several other technologies in this book, it is only the support of the very largest companies that will enable us to develop low-carbon alternatives at the speed the world needs. The chief executive of Renault-Nissan, Carlos Ghosn, once said, 'We must have zero-emission vehicles. Nothing else will prevent the world from exploding.' Ghosn has an apparently unshakeable faith in the necessity of a rapid swing away from the internal combustion engine. His partnerships with Better Place, his commitment to build a mass-market small electric car and his public pronouncements about battery technology have helped convince the world's motor industry that they too need to begin rapid product development of cars that make traditional large engines redundant.

Motor fuels
from cellulose

Second-generation biofuels

At the 1900 World's Fair in Paris, the Otto car company demonstrated an unmodified diesel engine running on peanut oil, not conventional fuel. Rudolf Diesel, the inventor of the engine that bears his name, noted the success of several other attempts to use food crops as alternative fuels and later wrote that 'power can … be produced from the heat of the sun, which is always available for agricultural purposes, even when all natural stores of solid and liquid fuels are exhausted'. One of the other great figures in the early history of the motor car had a similar view. In 1925, Henry Ford said:

> The fuel of the future is going to come from fruit like that sumach [a type of tree] out by the road, or from apples, weeds, sawdust – almost anything. There is fuel in every bit of vegetable matter that can be fermented. There's enough alcohol in one year's yield of an acre of potatoes to drive the machinery necessary to cultivate the fields for a hundred years.

Though few people realise it today, plant-based 'biofuels', as described by Diesel and Ford, were real competitors to petroleum fuels in the early decades of the motor car. Early Ford automobiles could run on alcohol and several US distilleries turned agricultural crops into fuel for cars until the 1930s. It was only the advent of

Prohibition that finally stopped the manufacture of alcohol as a fuel, clearing the way for the hegemony of crude oil.

The oil price shock of the early 1970s saw a temporary revival of interest in use of agricultural crops as a source of motor fuel. More recently, the US has strongly encouraged the use of corn as a source for gasoline. Government policy was initially driven by a wish to provide new markets for corn farmers, who had been suffering from declining incomes. It is only in the last four or five years that ethanol has been seen as a way of reducing the US dependence on oil imports and addressing the climate change problem. From a global warming perspective, food crops are potentially better than fossil fuels because burning ethanol simply returns carbon to the atmosphere that had previously been extracted by photosynthesis when the plant was growing.

Today many countries have policies to increase the use of crops of fuel. A large fraction of Brazil's petrol is made from sugar cane. Most crucially, about one in twenty of the world's cereal grains is now processed by US refineries into a gasoline substitute. To put this in context, over 100 million tonnes of North American corn is turned into biofuel each year, but this only shaves about 1 per cent off world oil demand.

Unsurprisingly, turning huge quantities of corn into fuel has tightened the world market for foods. Prices rose dramatically in 2007/8 and at the time of writing are still well above levels of five years ago. One International Monetary Fund survey indicated that the use of corn for biofuels was responsible for about 70 per cent of the increase in the world price for corn between 2004 and 2008. Using increasingly scarce food to make petrol has given biofuels a bad reputation. But advances in chemistry mean that we will soon be able to use agricultural wastes such as wood chips and straw to make petrol, just as Henry Ford forecast eighty years ago. The key is cellulose, a complex and tough molecule that forms a large part of almost every growing plant. Cellulose is the most abundant carbon-based molecule in the natural world, vastly more abundant than the simple sugars and carbohydrates we are now using to make biofuels. We need to find a way of cheaply and efficiently cracking the cellulose molecule and turning it into simple alcohol. There are challenges still to overcome, but with huge amounts of US venture capital flooding into the industry, cellulosic alcohol may well become the liquid fuel of the future. Combined with electric cars, cheap and environmentally benign ethanol can help slash carbon emissions from transport.

From bad ethanol to good

Add some yeast to sugary liquids in the absence of air and the resulting fermentation process will produce alcohol – called ethanol if you use it as a petrol substitute. Ethanol burns well and can be added to petrol as a supplementary fuel for today's cars. In fact, with some slight modifications, many cars on the road today can run perfectly well on almost pure ethanol mixed with just a small amount of petrol. In some ways, this fuel is actually better for cars than petrol. It delivers better acceleration and reduces the need for potentially dangerous additives. And although ethanol contains less energy in each litre than petrol does, new engines with high-compression cylinders may be able to turn slightly more of the fuel's energy into motive power than is the case with petrol cars.

Most of the biofuel sold in Europe and America today is made from foodstuffs, such as sugar cane, wheat and corn, which have been turned into simple sugars and then fermented by yeast. By contrast biodiesel is made by crushing seeds to capture the natural oils they contain. Diesel's engine exhibited at the Paris World's Fair used peanut oil, but today's favourites sources are tropical palm oil and rape-seed oil from temperate areas such as northern Europe. After harvesting, the oils are then put through a chemical process to create a diesel substitute. In both cases, high-value agricultural products are being diverted and taken through inefficient processes to create fuel for the ever-increasing number of cars.

Car owners filling up their tanks in Europe will be generally unaware that their expensive fuel already has a small amount of ethanol or biodiesel mixed with the fuel. As a means to get the industry started, retailers in some parts of the world are now obliged by law to incorporate biofuels into standard petrol. Similarly, diesel fuels have had plant oils added to the mix. When the large-scale move to biofuels began about five years ago, the world was a very different place. Agricultural surpluses were holding down the price of grains. Ethanol refineries that sprang up were a welcome addition to the flat landscapes of grain and corn producing areas around the globe. In Brazil, the sugar cane industry now produces huge quantities of the cheapest ethanol in the world from a crop that otherwise faced continuous downward price pressure because of the export subsidies of the rich world.

As we have begun to understand the true impact of ethanol and biodiesel production, however, serious doubts have arisen about the wisdom of obliging petrol companies to incorporate biofuels into their products. Most people now think that the benefits of first-generation food-based biofuels are outweighed by the problems they cause. Even the politicians who initially backed the ethanol industry are now edging towards the view that it would be mistaken to mandate that petrol should contain more ethanol. Food price increases in 2007/8 prompted riots and increasing worries about security of supply. Ethanol production also uses large quantities of water in areas in places where the supply may be already worryingly insufficient.

Furthermore, although ethanol made from crops may help reduce dependence on imported oil, it probably does very little to reduce the emissions of greenhouse gases. Growing wheat in Europe or corn in the American Midwest requires large inputs of fossil fuel energy to produce the fertiliser, look after the growing crop and process the grain into sugars and then ethanol. Moreover, when it breaks down chemically in the soil, artificial fertiliser produces a small amount of nitrous oxide, a greenhouse gas over 300 times more powerful than carbon dioxide.

Although the precise figure is the subject of fierce and bad-tempered disputes between scientists and fuel manufacturers, ethanol made in temperate countries probably saves less than 30 per cent of the greenhouse gases associated with a similar amount of petrol. As knowledge improves, we may find that ethanol from wheat and corn actually saves no emissions whatsoever. In particular, the evidence is growing that in some climates and soil types, growing wheat using nitrogenous fertiliser generates enough nitrous oxide to wipe out the greenhouse gas benefits of using wheat rather than petrol.

Despite this, European countries are still pressing ahead with a plan to ensure that at least 10 per cent of motor fuels are made from plant sources by 2012. Biofuels might conceivably reduce transport emissions by a few per cent in the EU, but this progress will be overwhelmed by the emissions from the extra cars on the road.

It also seems highly likely that biofuels exacerbate the problem of deforestation. Perhaps a fifth of man-made greenhouse gas emissions come from the clearing of forests. When a forest is destroyed, much of the carbon stored in its trees and soils becomes carbon dioxide in the atmosphere. As larger fractions of food production land are given over to ethanol and biodiesel, the pressure to cut down forests to replace the lost cropland increases. This is particularly important in the tropics. Old forests are being destroyed in order to plant oil palms for biodiesel in Asia. Even in Brazil the loss of the rainforest appears to be exacerbated by ethanol production from sugar cane. Although cane is produced in the drier parts of the country, well away from the Amazon, the use of agricultural land for growing crops for fuel is affecting the supply and demand for land across the entire country. The unpalatable conclusion is now almost undeniable: biofuels made from foodstuffs are adding additional stresses to an already overstretched world ecosystem.

The core problem is that the amount of energy used to drive people around is huge, far greater than the energy in the food that we eat. The average person in Europe uses about 700 litres of motor fuel a year, or about 2 litres a day. The figure in the US is over twice this level. We can easily calculate the amount of energy contained in this fuel and compare it to the calorific value of the food we eat. A food calorie is just another way of expressing a unit of energy. We can't create new energy by turning wheat or corn into petrol; this would break the laws of physics. All we can do is convert the energy into a different form. Every calorie we use to make motor fuel reduces the calories available to eat.

The result of the comparison between the energy used in a family car and the calories in food is very striking and somewhat depressing. The amount of energy Britons use to drive their cars each day is eight or ten times the calorific value of the food they eat. In the US, the multiple is about twentyfold. The comparison shows the folly of any attempt to use agricultural production as a means of reducing the greenhouse gas emissions from petrol. Even if we turned *all* our food into petrol, and lost no energy in the process, we would only produce a tiny fraction of our total need for motor fuel. There simply isn't enough viable cropland to feed 6 billion people and fuel hundreds of millions of cars as well.

Comparing the energy we use driving a car with the energy in the calories from the food that we eat is not strictly fair. Much agricultural land – some people say 60 per cent – is not actually used to produce food for human beings. Instead, it is devoted to growing food for animals, which we then eat. This is a very inefficient and wasteful process: an intensively farmed cow eats 8 kilograms of corn for every kilogram of weight in the slaughterhouse. If the world moved to a diet entirely composed of cereals, fruits and vegetables, then there would be huge amounts of surplus land no longer needed to produce food for animals. We could use these acres to make food to convert into petrol. But even if all this land was devoted to making ethanol feedstocks, the area could only provide a small fraction of today's consumption of fuel. Anyhow, the opposite is actually going on: as people get richer in the developing world, they are tending to adopt the dietary habits of rich countries and consuming more meat, thus increasing the total amount of land needed for food. The first-generation biofuels industry in the US contends that increased meat consumption in the newly prosperous countries of Asia has actually had more effect on food prices than converting corn into ethanol. In addition, more and more people have access to a car, so the amount of vehicle fuel needed will continue to rise for many years.

The conflict between food and fuel can be easily shown in an example. The most productive lands in the UK are largely given over to wheat. In a good year, these farms can produce more grain per field than anywhere else in the world, an average yield each summer of about 8 tonnes a hectare. Turned into ethanol in the most efficient processing plant in the world this might produce about 4,000 litres of ethanol, enough to cover the annual transport needs of about six Europeans. However, the food value in the wheat would give at least fifty individuals the calories that they need for a healthy diet. Unless we can substantially expand the area given over to crops across the world, ethanol from grains is in direct competition for the limited amount of land available for growing food.

Another problem is that the current technologies used to make ethanol from wheat and other foodstuffs are not very successful at turning the energy in the crop into liquid fuels. Large amounts of heat are needed to drive the process, reducing the net energy benefit of the grains. Paradoxically, it might actually be better to burn the crop and use the combustion process to drive turbines to make electricity. The energy from a tonne of corn would drive an electric car further than the same amount of grain converted to ethanol. In the long run, as Chapter 6 argues, electricity is a far better way of powering our motor vehicles. However, we have to live with the world as it now, with over 600 million cars on the world's roads, all but a few of which run on liquid fuels.

At some point soon, the tide may turn and politicians will start to campaign against *all* biofuels. This would be an unfortunate mistake. Biofuels made from the simple starchy and sugary molecules in food are just the first stage in the exploitation of biological materials for use as petrol and diesel replacements. The next generation of biofuels will not use the seeds of wheat and maize to make petrol replacements; they will use the much more complicated molecules contained in wood and agricultural wastes. Eventually it will probably be possible to process any complex material containing carbon atoms – plastics, municipal wastes, even the output from sewage farms – into a liquid fuel that can be burnt in a car engine. Henry Ford understood this almost a century ago. The technological problems are not especially large, but dicing large carbon-based molecules into simple alcohols such as ethanol is costly and difficult to achieve on an industrial scale. We know *how* to break down most hydrocarbon molecules using some combination of heat, pressure, water, acids and catalysts. But to make this process competitive with crude oil at a price of $70 a barrel, we need to be able to do it for less than 50 cents a litre at volumes of millions of litres a day. Therein lies the challenge.

Next-generation biofuels should enable us to avoid most of the problems that have arisen with wheat and maize. Many wastes, such as wood chips from sawmills, have few alternative uses, and their use for fuels will not increase the price of food. In addition, wood plantations do not generally use artificial fertilisers made from fossil fuels, nor do they require extensive cultivation from equipment burning large amounts of diesel. The same is true of grasses such as miscanthus and switchgrass, which are both very good sources of cellulose and can be grown on indifferent land. This means that the greenhouse gas emissions from cultivating the raw materials are low. In time, we should be able to produce liquid fuels that have a relatively low environmental cost, with accompanying greenhouse gas emissions perhaps 10 or 20 per cent of the impact of petrol or diesel. This is not to say that woody biofuels will not be found to produce other environmental problems of their own, but the evidence so far is that their side-effects are minimal compared to the burdens placed on the world by ethanol made from food.

How long will it take to get to the point at which biofuels made from wood, waste, and other unusual materials can compete on price with fossil fuels? The answer may be as little as five years, but it is impossible to be certain. It could be decades. As with other technologies in this book, progress to date has been slower than expected. Vinod Khosla, the legendary Silicon Valley venture capitalist, has invested in a wide range of US companies all trying to find low-cost, large-scale ways of breaking complex molecules into simple alcohol or other fuels. After a career in the computer software and networks industries, he has focused his almost limitless energy on technologies that may provide a way of cheaply converting cellulose to ethanol. But even he has found it slow going.

In an interview in January 2007, Khosla said 'this will be the year of cellulosic ethanol'. He was too optimistic. Although the stream of press releases announcing breakthroughs and cost reductions grew in volume throughout the year, the evidence of real progress was small. By March 2008, Khosla was telling the *Wall Street Journal* that 'the first commercial plants that are cheaper than both oil and corn ethanol are targeted to start operations at the end of next year [and will] probably be in full operation in 2010.' This probably won't happen either.

So the timings have slipped, and will probably slip further. But

Vinod Khosla, the man behind many cellulosic ethanol companies

there's little doubt that the engineering issues will be solved at some time in the next few years. The successful firm will achieve ethanol costs very similar to a litre of gasoline. One of the eight very different firms into which Khosla has put his money will get there, but even he cannot yet know which one will achieve this ambition first.

Of course, lowering cost isn't the only issue faced by the cellulosic ethanol industry. There's also the question of whether there's enough waste biological material available to fuel any sizeable proportion of the world's cars in twenty years' time. Khosla believes the answer is an unequivocal 'yes', though not everyone is equally optimistic and we'll discuss his analysis later in the chapter.

The previous chapter suggested that the world will eventually power most of its vehicles with electricity stored in rechargeable batteries. That is probably the most energy-efficient way of providing people with personal mobility. However, for cars still using internal combustion engines, including plug-in hybrids, we need to develop fuels that are not based on oil, coal or gas. A new car sold today will probably still be on the road in fifteen years' time, and so we will need petrol and diesel for at least that long, and probably many decades more. How will we get to the stage where we have substantial supplies of low-carbon ethanol which does not add to the pressure on the world's limited resources of good land?

In the US, first-generation corn-based ethanol is made using a two-step chemical process. First, finely ground cornmeal is added to water, and an enzyme converts the starches to dextrose, a simple sugar molecule. Then the liquid mash is transferred to a tank to which yeast is added. The yeast turns the dextrose into ethanol and carbon dioxide. Today, the most efficient ethanol refineries need about 2.5 kilograms of cornmeal to produce a litre of ethanol. At today's prices, this is more expensive in raw material costs than oil, and fermentation additionally requires large amounts of energy in the form of heat. So the US ethanol industry only exists because of legal requirements, import restrictions and extensive subsidies. The process is financially costly as well as being of minimal benefit to climate change. (Brazilian sugar cane ethanol is much cheaper to make because the raw material is already a simple sugar and because the heat for the process can be provided by burning wastes from the sugar cane plant itself.)

Corn and wheat have become expensive. By contrast, agricultural and forest wastes can be obtained for very little. The straw from fields, the leaves of corn cobs, and the chippings from a sawmill still have little or no monetary value. However, these materials, like all plant matter, contain cellulose. Straw and leaves are mostly composed of the molecule, and even in wood it can represent over 50 per cent of the weight, along with lignin and hemicellulose. Cotton has even more: about 90 per cent of its mass is cellulose. Our old clothes could eventually provide a feedstock for cellulosic ethanol refineries.

Plants and trees create large carbon-based molecules, like lignin and cellulose, to provide the structure and strength they need to grow and flourish. These compounds must be solid and tough to maintain cell walls and to provide the 'skeleton' of the organisms. It is therefore no accident that they resist the attempts of chemical engineers to break them into smaller units. Whereas corn starch needs relatively little encouragement to break into sugars and then into ethanol, cellulose is very stable. We can see this in the human digestive system: foods composed of starch are easily broken down, but cellulose, commonly known as roughage, passes through the body untouched by the fierce stomach acids and hungry gut bacteria. Cellulose-digesting animals, such as cows, sheep, and other

The structure of the cellulose molecule

ruminants, need much more complex digestive processes and enzyme-secreting bacteria to crack the cellulose molecule in their multiple stomachs.

Cellulose is a very long and straight chain of thousands of glucose molecules bonded together. An individual cellulose molecule binds strongly to its neighbours, giving cellulose its fibrous, rope-like characteristics. These complex structures have, of course, been made by living organisms from the simple ingredients of carbon dioxide and water using the energy made available by the photosynthesis process. So we can think of the usable energy contained in cellulose as stored solar radiation. This makes cellulose a source of renewable energy, but not necessarily carbon neutral: we may still need energy to grow and process the chemical energy stored in the tightly bound cellulose molecules.

There are scores of companies, almost all in the United States, trying to find the best way to turn cellulose into ethanol. They're seeking a process that's cheap, that can be carried out at a scale of hundreds of millions of litres a year, and that uses readily available sources of raw material, such as wood wastes. Most of the companies are focusing on one of two possible processes. The first option is to heat the cellulose in the absence of air to a very high temperature until it breaks into smaller molecules and eventually turns into simple gases such as hydrogen and carbon monoxide. These gases are then passed over catalysts to form ethanol, or bubbled through a stew of microbes that eat the dissolved gas and excrete ethanol. This is usually known as the 'thermochemical' approach.

The number of American companies racing to find the most profitable way to produce cellulosic ethanol probably even exceeds the large number of British companies trying to commercialise energy collection from the oceans. At the time of writing, it is impossible to tell which are going to succeed and which will lose all their investors' money. If past experience in other industries is any guide, we will see one or two manufacturing technologies emerge as the lowest cost way of making liquid fuels. When the winning approach becomes clear, most of the original young companies funded by venture capital will quietly disappear. At the same time, larger, less nimble businesses will enter the market, hoping to compete with the successful innovators who found the best way to make fuels cheaply.

Out in front at the moment is probably Range Fuels, one of the Khosla Ventures companies. Along with five others, Range won a $76m award from the US government to help it build its first commercial-scale plant and it has since had substantially more money from the taxpayer in the 2009 green stimulus package. The ground was broken in late 2007 for its new refinery in Soperton, about 150 miles from Atlanta, Georgia, in the middle of actively managed pine forests. The refinery will take the branches of the trees, which would otherwise be of little value, and use them to make what Range thinks will be the first commercially available cellulosic ethanol in the US. By the end of 2010, the company expects to have completed the building of the plant, a unit that will eventually turn out 20 million US gallons (about 75 million litres) of ethanol and other alcohols per year. When fully complete, the plant will refine about 100 million gallons (over 370 million litres) each year.

Early works at the Range Fuels Soperton refinery in the middle of the forest which will provide its feedstock

These numbers are impressive, but even this large refinery, costing over $200m, will ship less than 0.1 per cent of US gasoline demand. Future plants will probably be much less expensive, but these figures demonstrate the extraordinary scale of the investment that will be needed if cellulosic ethanol is to make a real dint in petrol consumption.

Mitch Mandich, who was CEO of Range Fuels during the initial phase of the company's development and is still a director, is upbeat about the company's technology. In 2007, when construction began on the Soperton plant, he announced that his company's process yielded ten units of energy for every one put in. This makes Range's approach one of the most energy-efficient of any of the cellulosic ethanol technologies now in development. If the plant turns out to achieve this, it will be an extraordinary improvement on corn ethanol plants, which may use three or four times as much energy to produce each litre of fuel.

To emphasise the wider environmental credentials of the Range Fuels approach, Mandich also said that the plant would only consume 25 per cent of the water used by a refinery using corn as a feedstock. He also described how the owners of the Georgia forests in which the Soperton facility is sited plant two trees for every one cut down, and how the refinery is engineered to produce virtually no waste products. If the refinery works as designed, the impact on the local environment will be very limited. This is another important way that cellulosic ethanol manufacture will be an improvement on first-generation biofuels.

The Range Fuels process is relatively simple. It uses the thermochemical route, converting solids to gases, and then the gases to liquids. The company has experimented successfully with many types of biomass, but at Soperton the feedstock will be unused wood from the forest. The material will come into the plant as fine woodchips. Heat, pressure and steam break the chips down in a process known as gasification. The resulting gases then react with steam to produce 'syngas', a mixture of carbon monoxide and hydrogen. (This is the same process used in many fuel cells, as described in Chapter 4.) After impurities are extracted, the syngas is fed into the second phase of the process: the transforming of the stream of mixed gases into ethanol and other liquids with similar molecular structures. This is achieved by passing the gas over a catalyst, an agent that induces a chemical reaction but remains unchanged itself. Further processing then maximises the overall yield of pure ethanol.

Which other companies look as though they might make it through the start-up phase? Zeachem is a Silicon Valley firm run by people with extensive experience in chemical engineering. The company's approach is to use a combination of chemical and biological processes. One stage uses a common bacterium that lives in termites and helps these wood-eating insects to digest their food. This microbe turns cellulose from wood chips into acetic acid, better known as vinegar. The acetic acid goes through an intermediate stage in the Zeachem process and is then turned into ethanol by adding hydrogen. The hydrogen comes from the lignin present alongside cellulose in wood and agricultural wastes. It is produced, as in the Range Fuels process, by gasifying the wood by applying intense heat. The energy needs for the process are provided by burning surplus hydrogen that it is not needed for ethanol production.

This process currently exists only in the laboratory, but Zeachem raised the funds in early 2009 for its first plant in Boardman, Oregon. Impressive claims for the technology include an extremely low cost of production – 80 cents a US gallon, or not much more than 10 pence a litre. Of course, this will only ever be achieved in a large plant when the technology has become mature. It is also dependent on finding extremely cheap bulk sources of woody material, so, as with all cellulosic ethanol producers, it's important to find raw material that the owner is willing to almost give away. For the initial plant Zeachem has signed an agreement with huge nearby farms that grow intensively managed and fast-growing poplar trees. Waste derived from processing the poplars will provide the feedstock. Indeed, all cellulosic ethanol producers are likely to site their plants near easily accessible forests or productive grasslands able to feed millions of tonnes of low-value biomass directly into the refineries with minimal transport costs.

Zeachem, one of the few visible cellulosic ethanol companies not part-funded by Khosla Ventures, is proposing a slightly more complex process than its competitors. If it succeeds, the three main advantages of its approach are likely to be that it achieves very high levels of energy productivity, that it does not create carbon dioxide as a by-product, and that it utilises almost all the products of the wood.

The claims for energy efficiency are impressive: energy output in the form of gasoline twelve times greater than the energy used to make the fuel, a figure that even exceeds forecasts for the Range Fuels' production process. However, the second advantage is perhaps even more interesting. Fermentation processes, such as those used to make corn ethanol, create large amounts of carbon dioxide. This is why there are bubbles in beer. To ensure that the global warming consequences of ethanol production are minimised, the gas must be collected and stored rather than vented to the open air. As the US and other countries move to mass production of replacement fuels, this is going to be a increasingly challenging task.

As importantly, a biochemical process that has carbon dioxide as one of the by-products has 'wasted' some of the carbon in the feedstock. Ideally we want all the carbon in the wood to be converted to a usable hydrocarbon fuel, not lost as carbon dioxide. A process that uses a chemical pathway that avoids an output of carbon dioxide, in this case by using bacteria that produce vinegar, has a real advantage over the conventional fermentation route. One additional implication is that this approach to manufacturing ethanol will produce very substantial volumes of fuel for each tonne of feedstock. Along with low process costs, and the ability to process millions of litres a day, a high yield from the raw materials is a vital characteristic of any technology hoping to displace petrol in the world's cars. Wood may be very cheap compared with wheat or corn – around £25 per tonne, as opposed to £100 or more – but it still makes financial sense to get as much ethanol out of it as possible.

The third advantage of the Zeachem process is that microbes eat the cellulose in the wood and the remaining lignin is gasified into hydrogen. Virtually nothing remains as a waste product. Although this is a feature claimed by many of the start-ups making ethanol from wood, Zeachem's claim has more plausibility than most. In a world where it is getting increasingly difficult to dispose of large quantities of any waste material, this is an important part of the attractions of moving to cellulosic ethanol. We can hope with some optimism that second-generation ethanol production will be a relatively clean process, with few serious environmental impacts.

Coskata, a company backed by General Motors, is also in the leading group of cellulosic start-ups. Coskata gasifies the raw biomass, producing a mixture of carbon monoxide and hydrogen. It then passes this mixture through a stew of microbes that consume the gases and excrete ethanol. This is relatively simple technology, and the company doesn't claim to achieve the high yields of Range Fuels or Zeachem. But its process can use a wide variety of fuels, including old tyres and municipal waste. Its refineries are also likely to be cheaper to build than for some of the other early-stage technologies.

Range Fuels, Zeachem and Coskata all propose to use 'thermo-chemical' processes to make ethanol. A second possible technique is to use enzymes to breach the defences of the tough cellulose molecule, creating much simpler sugars, such as glucose, and then use yeasts to ferment the sugars into ethanol. Industry insiders call this the 'saccharification' route, referring to the intermediate step of creating simple sugars. One major barrier to commercial progress on this road is the price of enzymes, which currently add at least 25 cents to the cost of producing a gallon of fuel. At the late-2009 oil price of $70 for a barrel, the cost of the crude used to make a gallon of standard gasoline is only about $1.65, so this is a major obstacle. Additionally, processes that use enzymes tend to need large amounts of heat to crack the cellulose open, a major additional cost. Once the cellulose has been turned into sugars, yeast is used to make the fuel, in a process very similar to making corn ethanol. This means that any cellulosic ethanol process that goes the saccharification route must have refinery costs at least as great as those involved in converting corn. Woody feedstocks are much cheaper than grain, so high processing costs at the refinery are not necessarily an overwhelmingly important problem. Nevertheless, huge efforts are being devoted to getting around these disadvantages by reducing the cost of enzymes and finding innovative ways to process sugars cheaply into ethanol. Without advances in these area, saccharification will probably be more costly than thermo-chemical processes.

Other routes are possible. One of the most interesting is turning cellulose directly into molecules very similar to those in petrol, rather than making ethanol, a petrol substitute. Such a fuel would have the advantage of being compatible with existing petrol pumps. It could also be used neat in the car's engine. Ethanol-fuelled cars, by contrast, tend to run best on a mixture containing 15 per cent fossil petrol, the presence of which inevitably reduces the carbon savings.

Businesses trying to commercialise these alternative technologies are running a couple of years behind the companies seeking to crack long-chain organic molecules to make simple alcohol, but some of their ideas show great promise. LS9, a company based in San Francisco, uses proprietary genetically modified bacteria to digest fatty acids from straw and other agricultural wastes. These bacteria then excrete an oily substance very similar to standard diesel. LS9 claims it can tweak genes in the bacteria to slightly alter the fuels that are produced. The company is targeting a fuel cost of about $50 a barrel, making it highly competitive with oil. It claims a net 85 per cent reduction in fossil fuels and points to another important fact: its products are far cleaner than conventional fossil fuel-derived products, meaning that carcinogens such as benzene are completely absent from the fuel and from the waste products of the production process. However, even the company says it will be at least 2011 before it will be in a position to make industrial quantities of its bacteria-produced fuels (although small quantities will be available in 2010).

Chemists can envisage several other pathways by which the cellulose molecule can be converted into simple motor fuels. We will probably see two or three further processes reaching the stage of pilot plants to test whether the chemistry works in commercial volumes. Perhaps $2bn or $3bn of private and public capital will be ventured on experiments with cellulosic ethanol technologies. This may seem a large amount, but in the context of the size of the market for petrol in the US alone – over $400bn a year – these sums are little more than small change. This should make us optimistic. The rewards for a successful company are so enormous that capital will continue to flow into the cellulosic ethanol industry until a solution is found to the relatively simple chemical engineering challenges discussed in this chapter.

My guess is that by the end of 2011 one or more producers will be refining a cellulose-based substitute for petrol at costs that are competitive with oil at prices as low as $40 a barrel. However, it is not just the chemical engineering that matters; equally critical is the question of whether the world has enough surplus biomass to make ethanol a serious competitor to petroleum-based fuels.

A cellulose bottleneck?

The crucial question for the proponents of alternative sources of petrol is this: are there sufficient amounts of unused plant matter, not useful as food, available to meet the gargantuan needs of the private car? Vinod Khosla is convinced there are. In a recent paper, he accepts that the land use issue is a serious challenge for ethanol. He says that the production of cellulose for conversion to petrol should only be encouraged if little or no additional land is required, so that the impact on food production is minimal. But the US needs over 1 billion tonnes of biomass a year to replace its gasoline use, even with optimistic assumptions about the yield that can be obtained by converting cellulosic material into fuel.

The scale of the task is enormous. The amount of biomass needed will almost certainly require about 80–100 million hectares to meet current US gasoline demand. This is an area larger than the farm-land devoted to crops today, and about the same space as occupied by US national forests, or the whole of Texas. If cellulosic ethanol is used to power fuel cells providing homes and offices with electri-city and heat, even more land would be needed. Weighed against this, the more electric cars there are on the road, the less severe the problem becomes.

The huge amounts of wood and waste needed for a petrol substi-tute do not deflect Khosla's optimism. He identifies three important sources of plant matter rich in cellulose that he thinks can provide the biomass required: winter cover crops, forest wastes and dedi-cated energy crops on marginal land not used for food.

Cover crops are used to maintain and improve soil structure. Planted after the main crop has been harvested, they can be left in the soil over winter. These crops are frost resistant and the green matter can be harvested in early spring. Khosla believes that cover crops might be able to provide 20 per cent of the total need for biomass. Land with cover crops can still be used to produce food in the summer, so there is no cut in food production. Unused wood from forests – wastes and trees that would otherwise have been simply left to rot – might add another 20 per cent, with most of the rest coming from crops grown exclusively for their energy value. In the US, the most appropriate energy crop is probably switchgrass, a perennial grass that grows happily on otherwise unproductive or even degraded land. In northern Europe, the most likely candidate for this role is miscanthus, a 4-metre-high oriental grass that pro-duces more weight of cellulosic material per hectare than any other crop in temperate latitudes.

Khosla carefully lays out his view that improvements in crop yield and in management of forests can produce enough cellulosic material every year. He also correctly points out that if all ethanol is made from woody wastes, the US will no longer need to divert corn from the food chain as it does at the moment. This important change will increase the amount of cropland available for crops. Since a quarter or more of the best US corn-producing acres are growing crops for turning into petrol, this is a major advantage. Cellulosic ethanol could therefore actually increase the amount of land available to grow food. Nevertheless, if Khosla's faith in the potential for cellulosic ethanol is correct, a large percentage of US land will still need to be employed to produce biomass for fuel. The incentives for land owners are probably substantial enough already. At 2009 oil prices, a biorefinery can probably afford to pay the $60 a tonne that Khosla says farmers will demand for their materials and still undercut the cost of fossil fuels. Khosla is also optimistic about future improvements in biomass yield, persuasively pointing out that while yields of food crops have doubled or tripled over the past decades, largely as a result of plant breeding and better agricultural practices, virtually no attention has yet been paid to making similar improvements for the plants and trees that will be used to produce biomass for cellulosic ethanol (or indeed wood for heat and power plants). The early results from experiments into breeding faster-growing trees have produced extremely successful results. There is very good reason to believe, therefore, that today's yields will be significantly improved, reducing the land area that will need to be given over to the new woody crops, perhaps by a factor of two or three.

Biomass such as dried grasses or wood chips is expensive to transport, not least because it is considerably less dense than coal or oil. For that reason, cellulosic ethanol refineries will be placed near to the land that provides their raw materials, whether forests or currently unused pastureland (in the Midwest, in the case of the US). The need to operate large refineries to ensure that operating costs are at their minimum means that each plant might need as much as 10,000 square kilometres (or an area of 100 kilometres by 100 kilometres) of land producing its feedstock.

This makes clear the scale of the task. If cellulose yields double, sufficient biomass to replace the US's gasoline demand of 600 billion litres a year will still require almost 50 million hectares, or half the land currently given over to crops. This is possible, but the

*Miscanthus, the super-fast-growing grass that may
one day be a common sight across Europe*

landscape around the world will look very different in thirty years' time as traditional slow growing trees and pastures are replaced with crops like switchgrass, or miscanthus in Europe and paulownia trees in the tropics. Khosla tells us to welcome this: it will provide poorer communities in the developing world with a good source of income and revive many of the depressed rural areas in the US and Europe. But as with the opportunity afforded by biochar (Chapter 9), land use changes around the world are going to be enormous, and potentially very unpopular. The few acres of tall miscanthus now growing in central England are widely disliked simply for being so different to the crops conventionally grown. When hundreds of square kilometres are given over to this tropical grass, we can expect much greater antagonism. But unless we decide to move very rapidly to electric cars which can be powered from renewable energy, we will need huge acreages to be devoted to the fastest growing energy crops, whether or not we like their appearance.

Of course, we can also hope to reduce the amount of fuel needed for each mile travelled. If engines are redesigned to run on ethanol, they will operate at higher compression ratios and fuel economy will be better. Smaller, lighter cars will also help. These improvements require manufacturers and legislators to aggressively support new technologies for improving fuel economy. Of course, the quicker the world moves to electric cars, the smaller the need for cellulose to be grown for conversion into ethanol.

Eventually, we will probably find that batteries are a better method of propelling cars. The typical driver makes very few long distance trips a year, and so, even if batteries continue to have limited storage capacity, the occasions on which people are going to be inconvenienced by needing to recharge en route are going to be limited in number. In the UK, the average person only makes twenty-eight car journeys a year of greater than twenty-five miles in length, less than 7 per cent of the total. Most people will be able to use electric cars. Commercial drivers may need to have cars that drive longer distances, and so continue to use liquid fuels, but these people are relatively few in number, although the distances they travel are far greater than the average.

We will also need cellulosic ethanol as an energy source for decentralised power plants, such as the fuel cells for office buildings and data centres discussed in Chapter 4. An office worker even in an energy-efficient building will typically need as much fuel for heat and power as needed for personal transport. Widespread use of renewable energy in fuel cells will inevitably require large acreages of land devoted to the production of material rich in cellulose.

The use of land for creating cellulose, or indeed the biochar discussed in Chapter 9, is going to be part of a much larger movement to implement agricultural practices that help maintain the carbon content of the soil. Conventional agriculture, both in developed and developing countries, lays little stress on the long-run maintenance of soil quality and the retaining of soil carbon. This attitude has to change, for reasons both of food productivity and climate change avoidance. The agricultural practices of fifty years' time will probably involve much greater use of crop rotation (the alternating of different crops so as not to deplete the soil) and the mixing of different crops in the same field, combining plants grown for their energy value with those grown for food.

The perfectly understandable push to increase food yields at almost any cost over the last few decades has produced monocultures that are highly susceptible to losses from disease and from pests. So although the move to very large-scale production of energy crops for making liquid fuels will involve substantial changes in land use, the world can probably cope without reducing the amount of food produced, provided we see substantial improvement in fuel economy, a switch to electric cars, and better agricultural practices. The trickiest question is probably not whether we can grow enough biomass to fuel our cars, but whether the world's agricultural land can both feed the poor and devote increasingly large amounts of primary food production to the fattening of meat animals as the global population gets more prosperous. There is no easy technological cure for the impact of the meat-eating habits of the rich world on the price and availability of nutrition for the poor.

Capturing carbon

Clean coal, algae and scrubbing the air

Power stations produce a large fraction of the world's carbon dioxide emissions. In developed countries, more than a third of the total greenhouse gas output typically enters the atmosphere from the smokestacks of fossil fuel power stations. Although capturing and storing this carbon dioxide is probably the single most important thing we could do to reduce emissions, power station operators have been slow to invest in research to show how carbon capture can be carried out economically. No one doubts the technical feasibility of separating carbon dioxide from the other gases and then storing it underground. The best example of this is the Sleipner gas field in the Norwegian North Sea, which separates the carbon dioxide and then stores it in an aquifer. However, it is a costly and complex process that needs to be replicated in thousands of power stations around the world.

Mention carbon capture to an environmentalist and the reaction will usually be unfavourable. Burning fossil fuel in a power station, collecting the carbon dioxide emissions, and then pumping them underground does not seem an ideal response to the need to reduce greenhouse gases. 'It just deals with the symptoms, rather than the causes' is a typical comment from climate change activists. Their view is that electricity should be generated from renewable sources and that the capture of carbon dioxide from coal or gas power stations is simply a means of delaying the much-needed switch to low-carbon sources of power.

However much one might sympathise with this opinion, carbon capture is going to play a vital role in tackling climate change.

World demand for electricity is increasing rapidly and the growth of renewable energy sources is simply not keeping up with the rate of growth. In other words, the percentage of electric power coming from fossil fuel sources is actually increasing today rather than decreasing, largely as a result of Chinese industrialisation and the easy availability of cheap coal in many parts of the world. Whether we like it or not, no successful attempt to cut global emissions can succeed without wide deployment of equipment to capture and store the emissions from existing and future power plants. Finding the right technology for coal power stations is particularly important: a unit of electricity generated from coal produces about twice as much carbon dioxide as a natural gas power station.

Carbon capture and storage, usually known as CCS, is the subject of intense interest among coal and electricity industries around the world. But, as yet, no working power plant has installed any form of large-scale CCS. The reason is simple. Capturing the carbon dioxide, liquefying it, and then transporting into safe long-term storage is expensive and technically difficult. A power station putting CCS equipment in place would be adding a substantial cost burden. Although the precise cost is not yet known, it is likely to work out at more than £20 for each tonne of carbon dioxide, adding over 2 pence to the cost of generating a kilowatt-hour of electricity, increasing coal generation costs by 40–50 per cent. Without a substantial and guaranteed financial incentive, no power station owner is likely to voluntarily move to CCS.

Forward-looking coal-fired power station operators are almost pleading with governments to ensure high carbon taxes in order to create such an incentive. Make carbon emissions costly enough and profit-maximising power stations will have an incentive to install capture equipment rather than pay for their carbon dioxide pollution.

'I am a carboholic', wrote David Crane in the *Washington Post*. Crane is the head of NRG, a US electricity generator with a portfolio of coal-fired stations. 'If the Congress puts in place a substantial carbon price,' he said, 'we will do what America does best; we will react to carbon dioxide price signals by innovating and commercializing technologies that avoid, prevent and remove

The Sleipner gas platform

carbon dioxide from the atmosphere.' He and other business leaders know that the technical obstacles facing power utilities are not insuperable, but until it becomes genuinely costly to emit carbon, utilities will drag their feet. The one major exception to this dilatory behaviour is the Swedish company Vattenfall, which is already investing significant sums. Its hugely significant pilot project is discussed later in this chapter.

The individual steps necessary before CCS is commercialised are well understood. We know how to get carbon dioxide out of a mixed stream of gases, how to compress it efficiently and then transport it, even over long distances. The Sleipner carbon capture equipment has separated tens of millions of tonnes of carbon dioxide from natural gas. Unusually, the natural gas contains about 9 per cent carbon dioxide when it comes out of the reservoir. This needs to be reduced to little more than 2 per cent before the gas is shipped onshore and sold to customers. Statoil achieves this reduction by passing the gas mixture through a liquid that absorbs the carbon dioxide but lets the natural gas bubble through. The absorbent liquid is then extracted and heated. The carbon dioxide boils off and the liquid can be reused. About a million tonnes of carbon dioxide a year are collected in this way and then re-injected into an adjacent aquifer. The reason it happened here first is largely because Norway already has a high carbon tax.

The Sleipner project demonstrates almost all the features that will be required in large-scale carbon capture at a power station. The exception is long-distance transport of the gas. However, the US already has a large carbon dioxide pipeline network for moving gas, demonstrating that this should be no obstacle to full-scale CCS. We just need to get power station operators to see it as in their long-term best interests to carry out the research necessary to combine these steps and then attach the carbon capture equipment to working power stations. The best possible encouragement is a high price for carbon emissions.

Capturing the carbon

You might think that carbon capture in a coal-fired power station would be relatively simple. Perhaps the carbon dioxide is separated off as it goes up the exhaust chimney and then pumped into a holding tank? Unfortunately it is not quite so easy. First of all, we need to understand a little bit about how coal power stations work.

Coal varies considerably in quality around the world but the basic technology for transforming it into electric power in most power stations is fairly uniform. Coal is pulverised into a very fine powder and then burnt in a stream of air. The combustion creates heat which then boils water and turns it into steam. This steam turns the turbines that generate electricity. Older power plants are only able to convert about a third of the heat value of coal into electricity but more modern power stations are designed to work at extremely high steam temperatures, which raises this efficiency to nearer 40 per cent. Burning the coal, which is mostly carbon, produces large amounts of carbon dioxide and other waste gases. Some of the other gases are severe pollutants and are removed from the exhaust stream. The carbon dioxide is almost invariably sent up a chimney where it escapes into the atmosphere.

Many of today's coal plants are antiquated. The average US plant is over thirty-five years old. They produce huge amounts of carbon dioxide compared with modern gas-fuelled plants, but because coal is relatively cheap these power stations are still economic to operate. Power station chimneys produce staggering quantities of gases. In terms of carbon dioxide alone, a single very big generating plant might produce 7 million tonnes each year, or nearly 1,000 tonnes an hour. At standard atmospheric pressure, this is almost 10,000 cubic metres a minute. In an old power station, this carbon dioxide will only account for around 10–15 per cent of the total exhaust gases. The remainder – perhaps as much as 10,000 tonnes an hour – is mostly nitrogen, which has passed untouched through the combustion process.

The simplest way of capturing the carbon dioxide from this mixture of gases is to bubble it through a solution of ammonia salts, much as Statoil does at the Sleipner gas platform in the North Sea. The carbon dioxide reacts with the ammonia compounds while the nitrogen floats upwards. The ammonia solution containing the dissolved carbon dioxide is then extracted and put into a large tank. It is mixed with very hot steam, which heats the solution and drives off fairly pure carbon dioxide. This gas can then be compressed, liquefied and sent to underground storage.

The major cost of this process arises from the large amount of valuable superheated steam that is needed to separate out the carbon dioxide from the ammonia compounds. This steam would have otherwise been employed in driving the generating turbines, so more coal has to be burnt to replace it. One recent study showed that 'post-combustion' separation and compression would reduce the percentage of the coal's energy turned into electricity in an old power station from 34 per cent to 25 per cent. Therefore the effect would be to reduce the amount of electricity generated from each tonne of coal by about a quarter. The cost of the extra equipment in which to carry out the separation of the carbon dioxide represents a substantial further burden.

The latest generation of coal plants, of which only a few have been built around the world, use two new approaches to generating electricity. The carbon capture process will also be somewhat different in each case. The first approach employs technologies similar to that used in gas power plants. The powdered coal is first heated intensely in a low-oxygen environment. This gasification process splits the coal into hydrogen and carbon monoxide (the 'syngas' described in Chapter 7). These combustible gases are then burnt in a gas turbine. Exhaust gases are used to raise steam which drives a second turbine, this time a conventional steam turbine. These new plants, inelegantly known as Integrated Gasification Combined Cycle (or IGCC) power stations, are more efficient at turning coal into electricity than older coal-fired power stations but also much more expensive to build. For our purposes, the important fact about IGCC units is that the operator can capture carbon dioxide more cheaply than in older types of coal plant.

In this type of power station, the carbon capture process will involve taking the carbon monoxide gas coming out of the gasification stage, prior to any combustion, and mixing it with extremely hot steam. The water molecules in the steam split into hydrogen and oxygen. The oxygen reacts with the carbon monoxide to form carbon dioxide, which is then extracted. This process is therefore a 'pre-combustion' carbon capture technology. The hydrogen from the steam is added to the hydrogen from the coal and the gas is burnt.

The main energy loss in this process arises from the need to heat the steam that oxidizes the carbon monoxide. Producing a clean stream of carbon dioxide in this type of plant will reduce the amount of electricity generated for each tonne of coal by perhaps 15–20 per cent. This extra cost is slightly less than for the other carbon capture processes. That's an advantage for this approach, though IGCC itself is a new and very expensive technology, not yet in widespread commercial use. The first plants have often disappointed their operators and it remains to be seen how long it will be before IGCC power stations become competitive with conventional plants.

This process for capturing the carbon dioxide is similar to how it might also be achieved in a gas-fuelled power station. BP investigated using the technique in a proposed new gas power plant in Peterhead, near Aberdeen in Scotland, but abandoned the plan when the British government announced that its early financial support for CCS would be entirely restricted to 'post-combustion' technologies and therefore that BP would have to bear the full cost of the CCS equipment and the higher costs of operating the plant. Since more coal than gas is used to fuel the world's power stations, and the volumes of carbon dioxide from coal are much higher, the British policy may have made sense.

There's a third possible approach, usually called the 'oxyfuel' process. If coal is combusted in pure oxygen, rather than air, the principal waste gas is carbon dioxide. There is no superfluous nitrogen. The power station therefore doesn't need to separate the carbon dioxide from the nitrogen after the coal has been burnt. After combustion, the only task is to extract the other pollutants and then compress and transport the exhaust gas.

This sounds a better solution, and may well be for some types of coal. But there is an obvious downside: it takes a lot of energy to produce the pure oxygen in the first place. Air has to be chilled to −200°C until it forms a liquid. The liquid air is then gradually warmed until the nitrogen boils off, leaving nearly pure liquid oxygen which is then extracted and allowed to turn back into a gas. Although coal burns better in almost pure oxygen, the net loss of energy is still substantial – almost as much as with the post-combustion approach.

Power stations of this type are still in the early stages of commercial development. But if a high carbon tax was introduced, we might see them widely rolled out. Importantly, oxyfuel equipment can also be retrofitted to existing coal power stations that currently burn the fuel in ordinary air. We do not yet know the cost or the likely impact on the amount of the coal consumed, but converting the older generation of power stations to the oxyfuel process may be a technology to watch.

The crucial point is that all these processes for capturing the carbon dioxide from a power station require large amounts of additional energy. This inevitably implies a cost penalty. Electricity produced by a plant with carbon capture is always going to be more expensive than that generated by a conventional power station. Unless legislation is introduced to mandate carbon capture, electricity generators will only switch to using CCS if the tax penalty for emitting carbon is higher than the cost of incremental energy used.

Storing the carbon

Compared with the challenge of capturing the carbon dioxide at the power station, processing and storing the gas is relatively simple. The carbon dioxide is compressed until it liquefies and is then sent by pipeline to the place where it is to be stored. Thus far, carbon dioxide has tended to be reinjected into gas and oil fields. It provides extra pressure, helping to push more of the oil and gas out of the reservoir. It is one of life's ironies that carbon capture at the power station could therefore result in compensating amounts of extra fossil fuels being burnt as a result of this extra production.

In the longer run, there isn't enough space in depleted fossil fuel reservoirs to hold the carbon dioxide from electricity generation. However, carbon dioxide can also be injected deep underground into saline aquifers composed of porous rocks, as it is at Sleipner. There is too much salt in the water in these reservoirs for it ever to be useful for drinking or irrigation, so little is lost by storing the unwanted gas there.

Although research on the subject is not yet conclusive, these underground rock formations could probably hold hundreds of years' worth of global carbon output. Usable aquifers exist under most of northern Europe, for example. The carbon dioxide from the Sleipner gas field is injected into the Utsira aquifer, which in itself may contain enough capacity to hold all Europe's power station emissions for centuries. When injected into these reservoirs, the carbon dioxide will dissolve in the water, forming carbonic acid. In some rock types, such as basalts, this will then combine with minerals in the rock to form very stable carbonates, effectively locking up the carbon dioxide forever.

Many environmentalists and policymakers worry about whether some of the carbon dioxide will eventually leak, returning to the atmosphere. The gas is buoyant and will try to escape upwards in an underground reservoir of any type. Some also talk of the risk of escaped carbon dioxide concentrating at ground level and asphyxiating living creatures. Or it might collect in groundwater near the surface, acidifying the water supply. The chances of a dreadful accident of this type are probably low, but the safety of the carbon dioxide storage process will depend on the exact conditions of the reservoir and its capping rocks. So far, there has been no evidence whatsoever that gas injected into the aquifer near Sleipner has bubbled back to the surface. We should be surprised if it did because the aquifer is covered by a thick layer of impermeable rock. Nevertheless, carbon dioxide leakage remains a concern and is one of the many aspects of CCS that needs urgent and comprehensive research. The rather glib assurances from some energy companies and others that the carbon dioxide 'cannot' escape are not sufficiently reassuring.

The cost of capture

How much will CCS cost? And will this cost be greater or less than the price of carbon? Will there be a legal obligation or financial incentive on power station operators to install carbon capture on all coal-burning plants? To electricity companies around the world, there are few more pressing questions than these. When a company invests in a new power station, it needs to be comfortable that the plant will work productively for several decades. Huge amounts of money are at stake. AEP, a huge American utility, is intending to spend over $2.2bn building an IGCC power station in West Virginia provided it can get approval from regulators. E.ON, the large German power generator, has outlined plans to put £1bn into a new coal-fired power station in Britain.

If CCS is added to plants such as these it will increase both the capital cost *and* the amount of increasingly expensive fuel it takes to generate a kilowatt-hour of electricity. So unless CCS is mandated by law, it will only make sense for power generators to incorporate the equipment if the price of carbon is high enough. The equation is simple, the carbon emitted from a power station needs to be more costly to the power station owner than the cost burden imposed by CCS. In autumn 2009, a power station operator in Europe faced a carbon dioxide price of about €14 a tonne. Up to this date, electricity companies have been given free allowances to cover their needs, but that doesn't mean there's no value in sequestering a tonne of carbon, thus reducing the power station's total carbon dioxide output. The company can then sell surplus permits in the European carbon marketplace, so reducing the carbon dioxide output by 1 tonne will add €14 to the balance sheet.

Is €14 enough to create the right incentive for power station operators? A new IGCC plant being built today with no carbon capture will generate about 700 grams of carbon dioxide per kilowatt-hour produced. At €14 a tonne, 700 grams costs about one euro cent. So if the carbon price stays the same, the sensible power station owner will invest in CCS if the extra coal needed and the other costs incurred add up to less than one euro cent per kilowatt-hour. An older plant might emit 20 per cent more carbon dioxide per kilowatt-hour, meaning that it could justify installing CCS equipment even with a slightly higher cost penalty.

As the section on the Vattenfall pilot plant later in this chapter shows, a €14 carbon price is not enough. It may need to be €30–40, sustained over several years, to get investors to take the risk and put in CCS equipment. The exact threshold for each power station will depend on the age of the plant, the type of coal used, and just how high a percentage of carbon dioxide is actually captured by the new equipment, but most analysts are broadly confident that CCS makes financial sense for a large fraction of European coal-fired power stations when (and if) carbon dioxide permits trade consistently above the €30–40 figure.

What about the US, which of course does not have a carbon tax? One suggestion from a group of energy companies, some of their major customers and a group of environmental organisations was that new power stations with CCS should be paid a premium for the electricity that they produce. The first 3 gigawatts of capacity (roughly two power stations) should be rewarded with $90 per megawatt hour. This incentive is high and would more than double the price of electricity from these plants but it would encourage risk-averse utilities to invest the capital to prove that CCS works. Later plants with CCS would receive a smaller premium, with the figure falling eventually to $30 a tonne, or not much more than the current European price for CO_2. This well-thought-through scheme is particularly encouraging because it shows that at least some of the generating companies that use coal for electricity production are confident that CCS can be made to work, and that eventually the cost will fall to moderate levels.

Gas-fired plants can also employ CCS, but the economics will be different, not least because a modern gas plant emits little more than half the carbon dioxide of an equivalent coal station. Nevertheless, even if CCS is used only for coal generation, the ultimate emissions savings from carbon capture will be enormous. Since in many countries coal-fired power stations represent over 30 per cent of total generating capacity, and in the US over 50 per cent, CCS is possibly the single most important technology opportunity described in this book.

Unfortunately, at current rates of progress at least another decade will pass before CCS can be shown to work across the wide variety of power station ages, burning technologies and coal types. The dilatoriness of policymakers and much of the energy industry on this issue has been deeply shocking. The need for an urgent programme of research has been obvious for five years or more but very little has actually happened.

As matters stand, it may be twenty-five years before most power stations are retrofitted with equipment to capture carbon. In other words, without a huge increase in commitment from government and power station operators, CCS is not going to be a quick solution to the climate change problem. As a result, many responsible people dismiss carbon capture, saying we should focus instead on renewables and on the reduction of electricity demand. This seems short-sighted. Coal is readily available in many parts of the world and is often the cheapest source of power station fuel. CCS is the best way we have of mitigating the impact of its use. Carbon capture techniques also have the considerable advantage of being exportable. We can only capture our own wind for electricity generation, but we can give China and India the technology to remove carbon dioxide from the growing number of coal power plants in these countries. (However, it is not entirely flippant to suggest that we might show how seriously we take climate change by sending these countries a large number of free wind turbines as an alternative today.) Of course, we cannot guarantee that these countries will use CCS, since it will always raise the cost of generating electricity, but a global carbon price should ensure that carbon capture eventually becomes in the financial interests of power station operators everywhere.

Electricity use is going to continue to rise. If the prediction in this book is right, we will eventually use electricity for much of our personal transport as well as for running appliances, lighting and machinery. Switching from inefficient internal combustion engines to battery-powered cars might increase the total demand for electricity by at least 10 per cent. We can hope that eventually much of the world's power will come from zero-carbon sources, but in the meantime it makes good sense to accelerate the R&D into carbon capture so that increases in electricity use don't simply result in more dirty coal-fired plants. Rising awareness of climate change issues, the threat of carbon pricing, and looming lawsuits in the US make power station operators in the rich world less likely to invest in coal generation than they once were. For example, many of the recent plans for new coal power plants in the US have been abandoned because of pressure from worried investors. Nevertheless, since so much existing power generation is coal-based, and because coal-fired plants are the easiest way of adding power to the grids of India and China, we urgently need to identify the cheapest means of capturing the carbon dioxide from this fuel.

One problem is that it's going to be difficult to incentivise the private sector into spending the large sums required to move CCS forward. Canny business people know that if their company were to find an improved method of carbon separation, they would stand little chance of making money from the new technique. Any substantial improvement would be almost immediately appropriated by governments (and with very good reason). When an invention is extraordinarily valuable, it is often impossible for the inventor to protect its ownership. Paradoxically, perhaps, it is the very importance of CCS that makes private sector research today so limited in scale. Carbon capture technology needs to be pushed by governments, perhaps by direct investment or the award of enormous prizes for specific and well-defined technical advances. Or, of course, governments could simply mandate the use of carbon capture in all coal power plants by a specified date.

The UK announced a carbon capture competition between power station operators. Hundreds of millions of pounds were to be made available to a company making the most impressive commitment to build a small-scale plant using CCS or to fit carbon capture on to a portion of an existing plant. It seemed a good idea but the government's offer of money to a single power station to build a demonstration plant has probably had the unfortunate effect of actually delaying research by several years. Rather than spend their own money, the power station owners have been waiting to see if they could get their R&D paid for. It would have been far better to fund a large prize that awarded several hundreds of millions of pounds to the first company that demonstrated the capture of 1 million tonnes of carbon dioxide. Then the power station operators would have had a real incentive to move quickly.

Vattenfall – the worldwide leader in carbon capture

Although most private power companies are notably uninterested in funding large CCS research projects – perhaps for the reasons offered above – the Swedish utility Vattenfall is one exception. The company is owned by the Swedish state so isn't obliged to maximise short-term rewards to private investors. The utility started research into viable forms of carbon capture in 2001, and began constructing its first pilot plant in 2006. This 30 megawatt power station is sited next to its existing coal power station at Schwarze Pumpe in eastern Germany. It started production in late 2008 and the plant will then run for ten years or more, experimenting with how best to capture carbon dioxide. Thirty megawatts is very small indeed by today's standards: E.ON's proposed new coal-fired plant at Kingsnorth on the Kent coast in southern England is over fifty times as big. But the German plant will be the world's first example of carbon capture at a working power station. The cost is about €70m, a substantial sum even for a major European utility.

Vattenfall says that oxyfuel combustion is likely to provide the cheapest way of collecting the carbon dioxide at Schwarze Pumpe. The process is a refinement of the oxyfuel carbon capture technique described above. Oxygen is separated from the nitrogen in the air onsite. Coal burns too readily if combusted in pure oxygen, so the temperature is damped down by reintroducing non-flammable carbon dioxide and water vapour from the exhaust stream. The rest of the exhaust is then cleaned of contaminants, such as sulphur

*The carbon capture plant at Schwarze Pumpe, with the
main power station visible in the background*

compounds, and cooled. When cool, water vapour condenses,
leaving almost pure carbon dioxide. Compressed to a liquid, the
carbon dioxide can then be safely stored.

Schwarze Pumpe burns lignite, a form of softer coal that produces even more carbon dioxide for every unit of electricity than
harder alternatives. Unusually for a power company, Vattenfall itself
mines the lignite locally. This makes it particularly appropriate that
Vattenfall has chosen to put its pilot carbon capture plant here. If
it wants to continue to extract and burn dirty lignite from the coal
fields of eastern Germany, it all too obviously needs progress in
carbon capture and storage.

Vattenfall plans to store the carbon dioxide in one of the rapidly
depleting fields in the Altmark gas-producing region about halfway

between Berlin and Hamburg. Pumping the carbon dioxide into the gas reservoir some distance away from the wellhead will help maintain the pressure of the relatively small amounts of gas left in these fields. This will increase the total volume that is recovered, adding just slightly to the climate change problem that CCS is meant to mitigate. The gas reservoir is about 3.5 kilometres below the surface and is overlain by several hundred metres of a highly impermeable rock, making it unlikely that significant amounts of carbon dioxide will ever escape. Nevertheless, local opposition is delaying the plans for the injection of the CO_2 into the field, even though this technique for recovering more oil has been used safely around the world for several decades.

The relatively small amount of carbon dioxide from the pilot plant at Schwarze Pumpe will be taken in road tankers to the Altmark gas field. The plan is that seven or eight tankers will cycle between the two locations, taking a total of about 100,000 tonnes a year. By 2016, Vattenfall plans to construct much larger CCS-equipped lignite power stations to further demonstrate the technology. If the reservoirs of the Altmark field prove suitable, the waste gas will probably be sent by pipeline from these second-generation plants. The total capacity of the whole gas field is likely to be over 500 million tonnes of carbon dioxide, meaning that if everything goes well, it may be able to accommodate all the carbon dioxide produced by a full-size power plant over the entire course of its working life. Other near-empty gas reservoirs in Germany might be able to hold the waste carbon dioxide from another four large power stations. But if carbon capture works and is employed at all coal-fired plants, deep saline aquifers will need to be employed, not just empty gas fields.

Understanding the lessons from the Schwarze Pumpe pilot plant will take several years. In the meantime, detailed planning will begin for the much larger demonstration plant in 2010, with the intention of producing electricity there by the end of 2016. The company estimates the cost to this point at over €1bn. If all goes well, an outline design for a full-size power station with CCS will be ready by 2020, almost twenty years after Vattenfall began its research. Two decades is a sobering length of time, particularly since most other large utilities have still barely started carbon capture feasibility studies.

Major design challenges remain but most people in the electricity business are quietly optimistic that CCS can work. Vattenfall confidently says in public that the technical problems are all solvable. 'It isn't rocket science', said a senior engineer at a recent conference. The company has also increased its own estimate of the proportion of a power station's emissions that could eventually be captured – from 95 to 98 per cent. Nevertheless, Vattenfall acknowledges that technical improvements will be needed to bring down the energy penalty. For example, it intends to drive water off the relatively wet lignite it uses in its German power stations. Despite the difficult future challenges, the company recently published an interview with JP Morgan's senior analyst covering European electricity companies. He said that on the basis of estimates provided by Vattenfall he expected that sequestering a tonne of carbon dioxide would cost as little as €30. By that reckoning, the hard-bitten board members of electricity companies would say that CCS doesn't make financial sense – yet. But Vattenfall itself has optimistically said that it eventually hopes to be able drive the cost down to below €20 when it learns the lessons from its early plants. Others aren't quite so optimistic and analysts talk of costs of €30–40 for several decades to come. Whatever the correct number turns out to be, it represents the single most important figure in the policymaking debate about how to decarbonise the world economy. If the eventual carbon price is significantly above this figure, we know that power station operators will have good financial reason to install CCS equipment and will do so voluntarily. Much below this level and we can be quite sure that they won't do it except under determined legislative attack. I think we can be certain that the quickest way to cut carbon emissions from the single largest source of emissions – the world's power stations – is to use the carrot of a high carbon price rather than the stick of legislation.

Governments are now beginning to realise that carbon capture offers significant prospects for carbon reduction, but that much research remains to be done. Vattenfall's commitment to the oxyfuel approach may be appropriate for lignite but conventional post-combustion techniques may be better for the bituminous coals that

An algae farm

are more commonly used in power stations. This is one of the many things we don't yet know. The correct answer may differ depending on the size of the plant, the space available, the local price of coal, and many other factors. It will be another twenty years before all these questions have been answered, but this isn't an argument for delaying research and development. When the history of the battle against climate change is written in a hundred years' time, Vattenfall's commitment to investing in carbon capture before it was commercially necessary will be seen as one of the most important single steps in the move to a low-carbon economy.

Other ways of capturing carbon

Capturing carbon dioxide using industrial equipment is all well and good, but it's not the only promising way of achieving carbon sequestration. Another approach is to feed the carbon dioxide from power stations, or simply from ambient air, to an unlikely environmental hero: algae. This idea is a form of 'biofixation', just like the techniques described in the following chapters to encourage the planet to store more carbon in its soils, plants and trees.

The main attraction of algae – a group of several thousand water-living organisms, ranging from large seaweeds to single-cell plants – is that they are extremely efficient at breathing in carbon dioxide. Most plants are not particularly good at this, as they only use 1 or 2 per cent of the light energy they receive from the sun to productively power the photosynthesis process. Some plants, such as maize, are better than others, but all land crops are very wasteful in the way they use light. Algae, by comparison, grow faster and capture more carbon dioxide. Under controlled cultivation, the weight of algae can more than double in a day. Provided light, water and nutrients, including carbon dioxide, are available, this exponential growth can continue forever. It is just possible that using algae to capture the carbon dioxide from a power station is a cheaper way of reducing emissions than all the expensive industrial processes described so far in this chapter.

But since the algae will all die eventually, and return the carbon to the air, how can these strange organisms help the climate change problem? The answer is that many types of algae have the additional advantage of turning some of the carbon from the air into usable oils. Under some conditions up to half the weight of certain types of algae is a form of vegetable oil. After extraction, this oil can undergo simple modification and then be used as fuel in standard diesel engines. The extraordinary fecundity of algae means that they create far more usable oil than conventional plants covering a similar area. One favourite industry statistic is that a hectare of algae ought to produce a hundred times as much biodiesel as a hectare of soy beans.

The implications of this are extremely attractive. A car that runs on biodiesel made from algae will be essentially carbon neutral from an emissions point of view. Yes, the action of burning the diesel will still result in carbon dioxide from the exhaust. But the gas will have previously been extracted from the atmosphere by the algae. In terms of the net effect on carbon dioxide levels, it would therefore be exactly equivalent to capturing the carbon dioxide from power stations and storing it underground.

In late 2007, the Anglo-Dutch oil company Shell invested in a new venture in Hawaii, where a start-up company is creating large tanks of open-air algae in coastal lagoons for eventual conversion into biodiesel. Unlike fossil diesel, this fuel contains no polluting sulphur and is harmless if spilt on the ground. No agricultural land is lost in its production, so the fuels will not reduce aggregate food production. In fact, algae may help us deal with the threat of long-run food shortages. The part of the algae that is not used to make diesel can be used as fodder for animals.

Shell is backing one way of growing algae: pools in the open sea. Algae can also be grown in inland ponds or in specialised 'bio-reactors' that keep the algae inside transparent plastic tubes. One company, Solazyne, is even intending to make batches of algae in the dark, creating growth through feeding the product with sugars rather than relying on photosynthesis. In all of these examples, the product can be harvested, dried, and then used for fuels and animal food.

Some companies look to go even further. They think that the best source of carbon dioxide for fertilising algae is actually the unmodified exhaust gases from coal and gas power stations. Flue gases, which as we've seen contain a maximum of about 15 per cent carbon dioxide, can be bubbled through water and algae. The organisms extract large amounts of the carbon dioxide to feed their growth and very little is left to be emitted to the open air. Could this be a cheaper and less energy-intensive way of separating the carbon dioxide from the harmless nitrogen coming out of coal-fired power stations?

The whole idea seems almost too good to be true and indeed the last few years have seen many false dawns for those who believe in algae as a means of capturing carbon from power stations. The entrepreneurs working to commercialise the 'biofixation' of carbon dioxide have faced setback after setback. One of the main problems has been that algae do not respond well to industrial cultivation. In large open ponds, controlling the water temperature is difficult and undesired species of algae can take over, reducing the useful yield. The growth process in enclosed bioreactors can also be difficult to control, and one famous large-scale experiment in 2007 saw excessive growth rates in the algae physically overwhelming the apparatus installed at a large power plant.

One of the many companies trying to succeed in harnessing the power of algae is A2BE, a company based in Colorado. A2BE has designed long tanks, enclosed in clear plastic, along which cylindrical rollers gently push the growing algae. One of the company's founders refers to the need 'to think like algae', in other words to understand that this green slime is part of the natural world and will not necessarily accommodate itself easily to man-made manufacturing processes. His business continues to have ambitious plans to deliver huge carbon reductions. He says that an area as small as 150,000 square kilometres (or 500 kilometres by 300 kilometres) would deliver a reduction of almost 4 billion tonnes of carbon dioxide a year, well over 10 per cent of today's total global emissions.

But will algae biofixation ever be successful on a large scale? There's no doubt that we can make algae grow, that this process absorbs carbon dioxide and that the oils contained in the organism can be extracted for fuel. What is uncertain is whether the process can be made economic on a large scale. Of course, this partly depends on the price of fossil fuels such as diesel. Major investments by the US Federal Government into research on algae were terminated decades ago because it looked as though the price of diesel from algae would never fall much below $3 a US gallon. That price is only a little more than what US consumers are paying at the pump and some of the scientists whose work was abruptly stopped twenty years ago are now back in demand as consultants to the universities and private companies furiously trying to overcome the problems they are facing with large-scale cultivation.

As with several of the technologies in this book, the case for large-scale and sustained research around the world is overwhelming.

Biodiesel from algae involves few, if any, of the problems of biofuels made from foodstuffs. It doesn't encourage deforestation, nor does it use a large amount of energy to grow and then refine. Its potential production rates per acre are a large multiple of what can be achieved with palm oil or any other tropical plant. It may be that instead of trying to develop algae on an industrial scale, with huge plants covering many square miles, we should try to farm it on a much smaller scale, using very simple equipment. This might mean lower yields per hectare, but it would allow farmers around the world to diversify a few of their acres into algae for use as biofuels and as an animal food, or even as a fertiliser for soil.

But we shouldn't give up easily on larger-scale algae plants. A quick look at the figures for the possible effect on carbon dioxide levels shows why. A square metre of water can grow 60 grams or more of algae a day if fed reasonable supplies of nutrients, including carbon dioxide. That means well over a half a tonne per hectare. The world uses about 80 million barrels of oil a day, and to completely replace all this crude with diesel fuel made from algae we would have to use about 30 million hectares, about 4 per cent of the area of Brazil or slightly more than the size of the United Kingdom.

This would be an enormous challenge, but perfectly possible, should the globe's leaders decide to focus on biofixation of carbon dioxide. The reduction in greenhouse gas emissions would be equivalent to at least 25 per cent of today's global total. And because most algae grow best in strong light and can cope with saline water, some farms could be placed in hot deserts with salty aquifers beneath them. The amount of water needed is not large and it can be recycled many times. The most efficient way of fertilising the growth is probably feeding the algae with the exhaust gases of power stations, though any source of carbon dioxide will do just as well. The other nutrients that the algae need include phosphorus, a mineral also needed to fertilise conventional farmland. Phosphorus can either be mined or, more sustainably, processed from human waste, which contains higher concentrations of phosphorus compounds than the rock extracted from most mines. In fact, human solid waste may turn out to be the best source, since mining phosphates is becoming increasingly expensive and difficult. The best possible locations for algae farms will therefore be next to power stations and close to sewage farms, land which for obvious reasons tends not to have high value in alternative uses.

It is too early to make a confident prediction but biodiesel made from crushed algae may turn out to be cheaper than ethanol made from cellulose. This would be a good outcome for the world since growing a tonne of algae will use far less land than a tonne of wood or grasses. Indeed, anybody wanting to bet on which technology will win Sir Richard Branson's $25m prize for removing a billion tonnes of carbon dioxide from the atmosphere might well consider a wager on sequestration by algae.

Biofixation of carbon dioxide doesn't have to employ algae. Horticulture can also be good at using carbon dioxide. On England's cool and cloudy north-east coast, 24 acres of greenhouses owned by supermarket supplier John Baarda grow tomatoes all the year round. The greenhouses are heated with waste heat from a nearby fertiliser plant, but the most important innovation lies in the use of the carbon dioxide that is also a waste product from the factory. Over 12,000 tonnes a year of high purity gas is pumped into the greenhouses instead of being vented to the air. This approximately doubles the ambient levels of carbon dioxide in the greenhouse atmosphere. The millions of tomato plants absorb the carbon dioxide through photosynthesis as part of their growth processes. This isn't really carbon sequestration because the carbon dioxide will return to the atmosphere when the plants die and the fruits get digested. But the 7,000 tonnes of tomatoes produced every year in this greenhouse complex are replacing fruit that would have been grown elsewhere, probably using much higher levels of artificial fertiliser. Since fertiliser production creates large amounts of greenhouse gases, horticultural reuse of carbon dioxide is an interesting and underexploited way of avoiding emissions.

John Baarda uses waste carbon dioxide from a nearby factory, but others are focusing on taking the gas directly from the atmosphere. This technique, picturesquely known as 'ambient scrubbing', requires carbon dioxide from the air to be captured and then stored or used. Carbon dioxide is only 0.04 per cent of the total volume of the atmosphere, so most people think that this makes little sense.

Surely, they say, it is easier to capture the much more concentrated carbon dioxide coming out of power stations. But Global Research Technologies (GRT) in Tucson, Arizona, believes it has found a way of cheaply and effectively capturing carbon dioxide directly from the air. The company has formulated a plastic that attracts and holds carbon dioxide molecules. When the strips of the plastic are fully loaded with carbon dioxide, they are placed in a humid atmosphere. The plastic also strongly attracts water molecules, which push the carbon dioxide away from the strips so that it can then be captured.

Though GRT was set up to focus on large-scale carbon capture and storage from the air, the company plans to demonstrate its approach and generate an initial income by producing products for the horticultural industry. One of the advocates of the technology – the eminent climatologist Wally Broecker – told me that the initial design is like a big waterwheel, half in and half out of a glasshouse and covered in strips of the plastic. As it rotates it picks up carbon dioxide in the dry external air. The strips enter the humid air inside the greenhouse and the water drives off the carbon dioxide, raising carbon dioxide concentrations in the air to perhaps twice the level in the air outside. The growing plants then capture the gas. Since carbon dioxide delivered in trucks to greenhouses costs over $100 a tonne, the wheel will pay for itself quickly.

After the technique is proven, GRT intends to produce units of the size of shipping containers that can collect at least a tonne of carbon dioxide each day from the atmosphere, to be sequestered underground, just like CCS at power stations. The indicative price of these units is currently about $100,000, meaning that at the late-2009 European price for carbon of about $20 per tonne, they would take about fourteen years to pay back their owners (or perhaps much longer if the cost of actually storing the captured CO_2 was taking into account). Clearly, substantial further cost reductions are necessary. The GRT machines would only take up a small fraction of 1 per cent of the land area of a densely populated country so they would use far less land than wind turbines that had a similar effect on atmospheric carbon dioxide concentrations. Additionally there will also be a need for a pipeline network for carbon dioxide so that the captured gas can be sequestered in saline aquifers or in the deep ocean, but this is no more difficult to engineer than the reinforcement of the electricity grids required to deal with higher levels of renewable energy production.

So there is a strong case for 'ambient scrubbing' devices, but we would need 2 million of these machines just to counterbalance the emissions of a single large European country. The cost for the UK alone might be nearly £100bn, or about 10 per cent of one year's GDP. Is this too much? It depends on how important you think it is to avert climate change compared to other objectives. The London Olympics will probably cost about £15bn when all the bills have come in, enough money to counterbalance nearly a sixth of the UK's emissions for a generation if it were spent on GRT's machines instead. It is still too early to say, but ambient scrubbing may turn out to be able to compete with CCS algae farming as a way of reducing concentrations of carbon dioxide in the atmosphere. Perhaps more importantly, if the world eventually panics at the sudden onset of obviously destructive climate change, building hundreds of millions of machines to actually take carbon dioxide out of the air may be the quickest way of beginning to reverse mankind's impact on the environment.

The next two chapters look at whether soils and forests represent attractive additional ways of collecting and storing carbon. The major advantage of using the biosphere in this way is that carbon contained in trees and soils does not need to be expensively transported in pipelines and pumped into aquifers. Nor does it need millions of large devices across the globe. In addition, if we use the land to store carbon, and we can find a way to reward landowners for their efforts, we will help to engineer greater involvement of developing countries in the global battle against climate change. Paying many of the world's poorest people $100 a year in return for helping us store carbon may be cheaper than any form of industrial carbon capture. We will also be compensating these people for inflicting climate change upon them. As I try to show, carbon capture in soils and through avoided deforestation will also help improve agricultural productivity. Equity demands that where possible we should prefer to use carbon dioxide reduction techniques that improve the living standards of the poorest people in the world, particularly those already suffering from diminishing rainfall and lower grain yields.

This is not an argument against carbon capture of emissions from power stations or the use of ambient scrubbing technologies such as GRTs. We need the widest possible portfolio of techniques for holding down emissions and then reducing carbon dioxide in the atmosphere. CCS at coal and gas power stations is a vital ingredient in any carbon reduction plan.

Biochar

Sequestering carbon as charcoal

Deep in the Amazon jungle are unusual patches of land where soil is darker and richer than in the rest of the region. These areas are highly fertile and also contain large quantities of carbon – carbon that has been drawn out of the atmosphere and safely locked away for hundreds or even thousands of years. These so-called 'terra preta' ('dark soil' in Portuguese) hold the key to one of the most exciting ideas in the fight against climate change. It's not about high-tech panels, turbines or vehicles. It's about rethinking the way the world uses a simple and familiar substance: charcoal.

The last chapter was largely about carbon capture using industrial processes. This chapter and its sequel deal with ways in which we can permanently store increased amounts of carbon dioxide in the world's soils and vegetation. Through the photosynthesis process plants and trees naturally take in carbon dioxide as they grow. Carbon from the air is used to make the complex sugar molecules that provide the physical structure of these living organisms. When they die, the carbon absorbed by the trees and plants generally returns to the atmosphere, either through burning or through gradual rotting. This process is often called 'the carbon cycle' and has been going on since the beginning of life on earth.

One way to beneficially disrupt the cycle is to part-combust wood to make charcoal, an almost pure and extremely stable form of carbon. This chemical stability means that unburnt charcoal sequesters carbon for centuries, even if it is simply mixed in with the soil. So, if we make charcoal from wood, and then dig it into the soil, we are sequestering carbon from the atmosphere just as much as if we were capturing it at a power station.

There are far too few quirky and unexpected ideas in the climate change field. Taking carbon dioxide from power stations to feed oily algae for making biodiesel is one such strikingly neat suggestion. Making charcoal, or 'biochar' as it is often called, is another. Previously sceptical scientists who have examined the impact of charcoal on the carbon cycle conclude that it does seem to permanently remove carbon dioxide from the air. Crucially, mixing charcoal in with soil has a very beneficial side-effect: it can significantly improve agricultural yields, particularly for topsoils deficient in carbon. As the evidence of biochar's effectiveness increases, research interest is growing around the world. This chapter looks at the impact of biochar added to arable soils, primarily in the tropics. The following chapter examines other ways of adding carbon to soils and forests.

Unlike many of the technologies discussed in this book, biochar doesn't necessarily need expensive equipment or highly skilled people. Capturing and processing the carbon dioxide from power station chimneys is technologically complex and expensive. By contrast, making charcoal is easy, and can be profitably done by poor farmers in the tropics.

It was the terra preta area of the Amazon that gave rise to the idea of adding charcoal to the world's soils to reduce carbon dioxide levels. Found spread all over the Amazon basin, these fertile patches are sometimes less than a hectare in extent, but in places they cover several square kilometres. Perhaps as much as 10 per cent of the region has soils with substantial amounts of added carbon. Terra preta exists in a wide variety of different soil types but they all share the characteristic high levels of charcoal residues. Research scientist Bruno Glaser from the University of Bayreuth in Germany says that a hectare of terra preta soil of a metre's depth typically holds 250 tonnes of carbon, compared with 100 tonnes in adjacent soils that have not been improved with charcoal.

The high fertility of the dark soils of the Amazon basin has been known for decades. Writers were commenting on it approvingly in the nineteenth century. Scientific attention first arose when the Dutch soil scientist Wim Sambroek published a highly influential book in 1966 on the man-made soils of Amazonia. He showed convincingly that the impressive soil productivity was a consequence of the high carbon content. Sambroek had grown up on a farm in the Netherlands, which had 'plaggen' soils – rich, deep and highly fertile. These soils had been created over centuries by local farmers adding thin carbon-rich turfs covered in cow slurry to the existing surface. He found that the Amazonian soils had similarly been adjusted by man to provide long-lasting improvements to their fertility.

With growing enthusiasm over the last few years, other researchers have noted that charcoal can improve agricultural productivity in many different types of soil around the world, doubling or even tripling yields in some circumstances and climatic conditions. We certainly do not completely understand the process by which biochar aids fertility. Much more investigative work needs to be done, but the evidence is strengthening that biochar's highly porous structure helps retain valuable nutrients and provides a protective structure that encourages the growth of beneficial microfungi. Look closely at a piece of barbecue charcoal, and you'll see thousands of tiny holes that were the cell walls of the original wood. This sponge-like structure gives a huge surface area that, it seems, helps make nutrients available and provides useful support to which beneficial organisms can cling. Increasingly, we also understand that in the dry tropics biochar assists in water retention and therefore aids crop growth, particularly in time of drought.

The structure of ground-up char may also help avoid the leaching away of valuable plant foods into streams and rivers. Biochar that has been laced with potassium, phosphorus or ammonia before being applied to the soil appears to achieve even better results than simple charcoal. In some ways we don't yet fully understand, the charcoal acts as a catalyst, making nutrients available to plants without being affected itself. Simply because it improves soil fertility, it may make good sense to apply biochar to a large percentage of the world's soils, even before we consider the potential impact on atmospheric carbon dioxide.

Laurens Rademakers, a Belgian social scientist with an interest in economic development in the tropics, set up an important trial in the West African country of Cameroon. For the two 2009

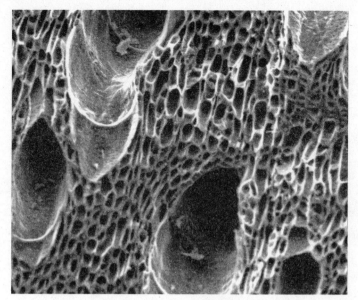

The tiny pores in a piece of biochar

corn-growing seasons, he persuaded many groups of subsistence farmers to apply large amounts of biochar to some of their plots and to compare the results with equivalent plots that hadn't been dosed. The biochar was made from the agricultural wastes of a previous crop. The waste would usually be burnt in open fires to dispose of it so making biochar does not divert organic matter from other uses. So far, the results of the trial have been impressive demonstrations of the ability of biochar to reinvigorate degraded and vulnerable croplands. On those plots that have had the highest applications of the charcoal, adding the biochar approximately doubled food yields. This means that biochar had as much effect as adding large amounts of expensive fertiliser. When fertiliser (organic and mineral) was also added, the yields were even better. Rademakers is far too good a scientist to claim that one set of trials, albeit a large and well designed one, is enough to prove that biochar works, but the improvement in yields certainly supports a hypothesis that it has a beneficial effect and provides a use for wastes that otherwise would simply be burnt.

Big problems remain to be overcome. The first is how to move from a handful of small-scale plants making biochar to exploiting the idea all around the world and sequestering billions of tonnes of carbon each year. The second is to find out why biochar doesn't invariably add to soil productivity. Occasionally, fertility may even be adversely affected. No one has yet worked out why. We will need much more research on what plant materials should be used to make the biochar, the right temperature for the charcoal kiln, and what other fertilisers should be added to the soil as the same time, perhaps already bound to the biochar. But these are questions which science should be able to solve without too much difficulty. Biochar stands a good prospect of being one of the simplest, cheapest and most effective ways of capturing carbon dioxide from the atmosphere and storing it safely.

Biochar at the smallest scale

Irishman Rob Flanagan is one of a growing network of practical idealists around the world working on biochar. His life was changed when he saw a 2002 science programme on BBC television about the terra preta soils of the Amazon. Excited by the visible evidence that biochar could help apparently infertile soils support productive agriculture, Rob went to work for EPRIDA, a pioneering biochar company just outside Atlanta in the US state of Georgia. After a couple of years working there, learning how to make charcoal, he carried out his research in tropical China and in Indonesia.

Biochar can be made in tiny kilns that double as highly efficient cooking stoves or it can be made in huge chambers that produce hundreds of tonnes a day. Flanagan's interest is in biochar at the smallest scale. His mission is to design a simple, cheap and reliable domestic cooking stove that uses locally available materials for its fuel and, as a by-product, gives the householder some charcoal to feed the soil. Some – perhaps a large part – of the world's deforestation is being caused by families seeking wood for cooking. Rob wants tropical households to have fuel-efficient stoves which could reduce the amount of wood needed, sequester carbon *and* boost soil fertility.

The smoky and inefficient open stoves in use in much of the developing world are highly wasteful of wood, increasing the

Rob Flanagan's prototype biochar stove

amount which needs to be cut down, and raising the amount of carbon dioxide released into the atmosphere, as well as reducing air quality in and out of homes. About 1.4 billion tonnes of wood and other organic matter is used for cooking fuel each year. If rolled out universally, efficient domestic biochar stoves, such as those being developed by Rob Flanagan, might reduce this figure by a half or more. This would avoid the need to cut down virgin forest, potentially slashing global carbon emissions by 10 per cent.

Importantly, Rob Flanagan's stoves can also burn agricultural wastes, such as rice husks, which would otherwise be unused. His stoves leave a good percentage – perhaps 15 per cent by weight – of the original fuel as charcoal in the bottom of the stove. This charcoal could be burnt as fuel but Rob believes that the best use is as a high-quality soil conditioner and fertiliser.

In one of his recent experiments, Rob compared the germination and growth rates of a fast-growing native Chinese tropical tree in seed trays with and without biochar. The differences were dramatic. The soil fertilised with small quantities of biochar produced much more vigorous plants. The leaves were a healthy dark green. The roots were stronger, too, and it seems likely that plants with this preparation will grow much more rapidly when transplanted to their final growing location.

To Rob Flanagan, this suggests a virtuous circle. The charcoal in the soil helps new trees to grow rapidly, which increases the amount of available wood fuel and, therefore, the volume of future charcoal. He wants to see biochar stoves in every agricultural village in Asia, reducing the amount of wood used and decreasing the amounts of money spent by poor families on fuel. Billions of new trees are being planted by the authorities in regions under threat of desertification, but some Chinese regions are still experiencing a loss of forest cover. This increases the carbon dioxide going into the atmosphere and affects the local climate. The absence of respiration (the return of water vapour from trees to the atmosphere) is increasing the threat of drought.

Equally importantly, Flanagan's stoves burn extremely cleanly, improving the air quality in homes. The World Health Organisation reports that a million and a half people die every year from the effects of indoor air pollution which is mostly caused by smoke from open fires in poorer communities.

Like the other experimental scientists working in the field, Flanagan isn't sure why biochar adds to the soil's fertility. When I chatted to him by email, he described this as the 'million-dollar question'. The fertilisation effects of charcoal are clearest in the tropics, but many researchers are now seeing similar improvements in the soils of temperate lands. Rob has seen extremely good results in New

Zealand, for example. But he admits that much research will be needed to work out the optimal temperature for his stoves and what types of woody fuels make the best charcoal for the soil.

It is still a frustrating time for the dedicated biochar researchers working in the tropics. They all strongly suspect that biochar has a very important role in climate change mitigation, as well as in improving living standards in poor communities. Research funding has been difficult to obtain, however, and people like Rob Flanagan are working on isolated experiments that should be happening everywhere around the globe. Findings are shared via some very active email discussion groups, while short videos giving the results of the latest experiments are uploaded to YouTube. But despite the remarkable effect on some plants and the huge potential for carbon sequestration, he still finds initial scepticism everywhere. 'It all just sounds too easy,' he says, 'no wonder no one gets it.'

Rob is trying to develop a very low-cost stove that will be easy to use and which local people will want to use for their cooking. As many other attempts to improve cooking practice around the world have shown, even the best stoves require some skill to operate. The women who do most of the world's cooking will need to be convinced that Flanagan's revolutionary stove will work. There's also the question of price. When built in large volumes, the stoves should cost substantially less than $40, but this is still an enormous sum for the very poorest people to afford. Nevertheless, a stove should repay itself quickly by substantially reducing the amount of fuel that the household has to buy or collect. We can hope that microfinance banks, such as Grameen in Bangladesh, will be able to lend money for biochar stoves because of the long-run savings to the household budget. Similarly, the growing number of institutions offering voluntary carbon offsets for people in the rich world might choose to subsidise the sales of Rob's stoves in the poorest countries. There are fewer projects where the carbon savings are more obvious.

But getting interest from the international bodies that concern themselves with climate change is difficult. Unlike some of the technologies covered in this book, the opportunities for big international companies to build large businesses based on sophisticated technology are simply not present. Large pools of US venture

BEST Energies prototype plant

capital are flowing freely into solar photovoltaic technology and second-generation ethanol manufacture. But biochar, an approach to carbon sequestration that is potentially also hugely beneficial to tropical agriculture and the deforestation problem, is virtually ignored. The same problem arises over research and development of drugs for many tropical diseases. Although malaria debilitates and kills millions of people a year, large companies cannot see substantial future profits and are unwilling to invest as much as in drugs that will be bought by wealthy societies. With small biochar stoves, the low income of most of the likely beneficiaries holds back research, development, manufacture and marketing.

Scaling up

Rob Flanagan's focus is small-scale biochar production in poor rural communities. BEST Energies in New South Wales, Australia, operates at the other end of the spectrum. BEST has spent the last decade developing a technology that can be scaled up to part-combust almost 96 tonnes of dry biomass each day, generating perhaps 30 or 40 tonnes of biochar. The biomass doesn't have to be wood. It can be agricultural waste or even municipal garbage.

The biochar from BEST's pilot plant is added to Australian soils, which have some of the lowest carbon levels in the world, and the results have been spectacular. Dr Lukas Van Zwieten, a scientist working for the New South Wales government, found that applications of 10 tonnes of biochar per hectare tripled the mass of wheat crops and doubled that of soybeans. Even if only part of this improvement is seen in other experiments, biochar would revolutionise the food production potential of carbon-starved soils. Laurence Rademakers' work in Cameroon has yielded similar improvements.

When woody wastes are heated to high temperatures, the material gives off gases and material that will condense back into liquids. These gases and liquids can be burnt directly as fuel. The BEST biochar plant will also produce significant amounts of useful heat from combusting the by-products in this way. Hot gases can be fed into a turbine which generates electricity. As coal and gas get more expensive, using biomass to make electricity is looking more and more attractive. Chapter 4 described how a similar technology that completely gasifies beetle-riddled wood is being used as a source of heat and power in western Canada. BEST Energies' process keeps some portion of the wood as charcoal.

EPRIDA, the US business that gave Irishman Rob Flanagan his first direct experience of biochar, has similar ambitions to those of BEST. It wants to build large-scale processing units that use woody wastes to create electricity and biochar. The difference is that EPRIDA believes it can make the biochar even more productive by adding fertilisers into the char as it is produced. The company's process takes the hydrogen that comes out of the wood and turns into a slow-release ammonia-based fertiliser bound into the pores of the biochar. EPRIDA says that this encourages the growth of a particular type of fungus that lives on the roots of most plants. These fungi feed using extremely thin hair-like tubes that gradually reach inside the pores of the biochar. The tiny tubes exude a protective glue-like substance that also binds together tiny bits of dead organic matter. This glue is very chemically stable and helps maintain the soil structure and its overall carbon content in addition to boosting plant growth.

How much difference could plants such as those proposed by BEST and EPRIDA make to carbon emission? This partly depends, of course, on whether biochar adds to the pressure to cut down the world's forests. Sequestering carbon in the soil, however beneficial for agriculture, is not going to reduce atmospheric carbon if the charcoal makers cut down virgin forest to make the biochar. That caveat aside, biochar plants could make an enormous difference. For the sake of illustration, let's say that Germany, Europe's largest emitter, decided to focus on biochar as a way to take its emissions to zero. Plants such as the ones offered by BEST Energies can produce 30–40 tonnes of biochar per day, so we would need about 20,000 machines to cut emissions by 250 million tonnes of carbon a year – more than enough to wipe out the country's total greenhouse emissions.

If we created the right economic incentives, it would be perfectly conceivable to install this number over a five-year period at a rate of 4,000 plants a year. For comparison, there are also about 20,000 wind turbines in Germany, most of which have been erected in the last few years. The crucial signal that entrepreneurs need in Europe or in a poor country reliant on agriculture is a high and stable price for sequestering carbon. While many countries now reward companies in the form of grants and subsidies for producing low-carbon electricity, no government pays money to businesses putting carbon back into the soil, an equivalent activity in terms of its effect on carbon dioxide levels in the atmosphere. As a result, commercial biochar companies are currently struggling to get the early commercial orders that will prove their technology works outside research laboratories.

Local expertise

Rob Flanagan's domestic stoves and BEST's large-scale plants are at the extreme ends of biochar production techniques. A middle way is to use existing local expertise in charcoal making. The only

A charcoal rick in Germany around 1900

difference from what already happens today is that more would be produced and, instead of all the charcoal being used for fuel, some part would be diverted and dug into the soil.

Charcoal-making is carried out in countless different ways around the world, but all the methods have something in common. They restrict the flow of air into a pile of burning wood or plant matter such as coconut husks. This keeps the temperature down and the lack of oxygen stops much of the carbon burning. At the relatively low temperature in charcoal kilns, volatile gases and liquids are driven from the woody material, and these can be collected or burnt either for heat for cooking (as in the case of a Rob Flanagan's stoves) or electricity generation (in industrial-scale plants). The liquid driven off, often called 'wood vinegar', is sometimes separated and used for a variety of purposes, such as an insecticide or even as a health drink in Japan.

Charcoal is what remains when all the volatile matter has been driven off. Because it is almost pure carbon, charcoal itself is an excellent fuel, burning cleanly, evenly and at a high temperature.

There is a considerable skill involved in making charcoal. Throughout history, communities have tended to employ specialist charcoal makers who earned their living making fuel for their neighbours. The charcoal maker needs to know how wet or dry the wood should be, how to limit the flow of air into the wood and when to open up the burning pile to collect the valuable charcoal.

Some cultures make their charcoal by building a pile of wood and then covering it with dense and non-flammable matter to reduce the airflow. In many place in northern Europe, for example, charcoal is made by covering a carefully made log structure with small branches before completing the pile by adding turfs to stop too much air getting to the wood. Other societies use purpose-built buildings while some make their charcoal in pits in the ground. The woody material is usually set alight by inserting a flame into a chamber in the middle of the pile. The pile may take days to turn into charcoal and during this time it must be carefully watched to ensure that it does not combust too freely. When the charcoal is ready, the pile is broken open and the hot charcoal is dampened with water to stop it burning further.

When farmers and small businesses make charcoal today, they are not doing so to dig it into the soil. They produce charcoal because of its value as a fuel. In the developed world, charcoal is mainly associated with summer barbecues, but in less well-off countries it is valued as a light and clean indoor fuel. In some regions it is the most important source of energy for cooking. Made in the countryside, charcoal is easy to handle and, being lightweight, easy to transport to nearby cities. In historical times, it was also widely used to smelt ores for metal. For example, charcoal will burn at a sufficiently hot temperature to create liquid iron from iron ores. No doubt in some parts of the world it is still used for this purpose.

So charcoal making is established and well understood across the globe. If we are to use biochar to help mitigate climate change, we need to massively expand the amount of charcoal being produced in poorer rural areas. But because the skills already exist in such communities, this is not an impossible dream. The crucial issue we face is finding a mechanism that rewards village-level enterprises in developing countries for producing charcoal and adding it to soils.

By any standard, the cost of storing carbon in the form of biochar is extremely low. In poor countries many people earn very little money for their work – perhaps just a dollar a day. Charcoal makers producing biochar in larger kilns in an agricultural community might be able to make several hundred tonnes every year. If they were paid for their charcoal even at the current European market price for carbon dioxide, they would earn wages that they could never hope to make in other occupations. The equipment that they need is simple and available locally and so there are no obvious obstacles to building large charcoal industries even in very poor countries.

Importantly, and perhaps surprisingly, verifying that carbon has actually been stored in soils is not difficult. Soil carbon levels can be measured simply and accurately with cheap equipment. New techniques even allow remote sensing of carbon levels using satellites. Of course, taking such measurements in millions of fields around the world each year, and remunerating the relevant farmers, will present enormous challenges. Nevertheless, the scope for carbon sequestration using biochar made in poor rural areas is huge. If organised on a sufficient scale, it may be no more expensive than the complex engineering required for carbon capture at coal-burning power plants.

The importance of soil carbon

We generally focus too much on carbon dioxide in the atmosphere. The world's soils hold more carbon than the atmosphere, trees and plants put together. The thin layer of soil which covers much of the earth's surface, but is rarely much more than a metre thick, produces a very large fraction of our food and sustains life as we know it.

Where is the world's carbon?

Plants and trees	600 billion tonnes
Atmosphere	800 billion tonnes
Organic carbon in the soil	1,600 billion tonnes
Fossil fuels in the ground	4,000 billion tonnes
Oceans	40,000 billion tonnes

And for comparison:

Carbon added annually to the atmosphere from fossil fuels	8 billion tonnes

The location of the world's carbon is not static, of course. Carbon in its various forms moves from soil to air to plant and in the reverse direction all the time. The amount of carbon – usually in the form of carbon dioxide – flowing in and out of soils in the natural cycle is perhaps ten or twenty times the volume put in the atmosphere by the burning of fossil fuels. In other words, we don't have to disrupt this cycle by a large percentage in order to achieve real reductions to the amount of carbon dioxide ending up in the atmosphere.

One further comparison should strike us forcefully. The weight of carbon in the soils is about 1,600 billion tonnes. Mankind's actions produce about 8 billion tonnes of carbon in greenhouse gases every year, of which about half is added to the atmosphere (causing climate change), with the rest absorbed by oceans, plants and soil. As the planet warms, one side-effect could be a small annual reduction in the soil's ability to retain carbon. As temperatures rise, many chemical reactions in the soil will speed up, meaning that carbon dioxide, an important end-product of many of these reactions, will tend to escape more rapidly back into the atmosphere. This is one of the many 'tipping points' that may threaten to produce a speeding up in the pace of climate change.

Worryingly, an increased flow of carbon dioxide from soil to air may be happening already. One recent paper in the science journal *Nature* suggested that the UK's soils are losing carbon at the rate of 0.6 per cent a year. The loss in the most carbon-rich British soils is over twice this rate. It isn't yet clear if the same is happening across the rest of the world, but as temperatures increase and rainfall becomes more erratic in many areas, there are good reasons to assume that there will be a net shift of carbon from soil to air.

A 1 per cent loss of the carbon in the soil across the globe would approximately triple mankind's total emissions for a year. We know so little about the impact of changing temperature and rainfall on the world's soil that this simple comparison should make us extremely concerned. By burning huge weights of carbon-based fuels and raising the equilibrium temperature of the atmosphere, we are, as is so often said, conducting a gigantic and dangerous experiment on the planet.

Modern agricultural techniques are another reason why soil

carbon levels may be falling. Ploughing, overuse and erosion by wind and water all tend to reduce the carbon content of the land. One respected researcher says that some agricultural soils have lost 50–70 per cent of their embedded carbon content since the Industrial Revolution. As more of the world's land is pressed into intensive agricultural use, soil degradation is probably becoming an important source of carbon emissions into the air.

A third cause of the loss of soil carbon is deforestation. When a hectare of trees is cut down, perhaps because a farmer wants to convert the land to agriculture to feed the world's growing population, the carbon in the wood is lost, usually in a fire used to clear the land for cultivation. Flushed out by rainfall, the newly exposed soils also rapidly lose much of their embedded carbon.

The last half century has seen truly remarkable increases in agricultural productivity. Average grain yields have almost tripled. Despite the rapid growth of the world's population, a smaller percentage of the world goes to bed hungry than ever before in history. But even to some optimists, this improvement shows disturbing signs of tailing off, with the dramatic recent food price rises being a sign of things to come. The supply of new land suitable for arable farming looks surprisingly limited. The Malthusian trap – food production growing less fast than population – is a spectre that many commentators laughed at a few years ago but now frightens increasing numbers of policymakers. One respected US government forecast shows the growth rate of cereal production dipping below the rate of increase in global population within a few years. Rising meat consumption will compound this effect as more and more grain is diverted to feed cattle and pigs.

So if biochar did nothing else but add to the productivity of our soils, help stabilise soil carbon, and reduce the pressure to cut down forests, we would have good reason to encourage its manufacture. But done on a sufficiently large scale, it could also make a real difference to the amount of carbon dioxide in the atmosphere.

Biochar's carbon impact

At the moment, carbon dioxide levels in the atmosphere are rising by about 2 parts per million (ppm) every year. This rate is increasing, partly because fossil fuel use is growing, and partly because the traditional stores of carbon – soils, vegetation and the oceans – seem to be becoming less effective at soaking up a large slice of man-made emissions. Such evidence as we have suggests that a slightly larger fraction of the carbon dioxide generated by our actions each year is ending up in the atmosphere. In view of this, what should be our target for biochar? Perhaps optimistically, we might look for it to reduce atmospheric concentrations of carbon dioxide by 0.2 ppm each year, about 10 per cent of the current net increase. That would be equivalent to eliminating the total emissions of the UK, France and Germany.

To achieve a reduction of this scale, we'd need to create about 400 million extra tonnes of charcoal and sequester it in soils. Averaged across all types of organic material – wood, plants, shrubs and grasses – biomass is about 50 per cent carbon. So if we could convert all of that carbon into biochar we'd need around 800 million tonnes of source material. In reality, the process would probably be only around 50 per cent efficient, so we'd actually need to process about 1.6 billion tonnes.

Processing such a huge amount of vegetation would be a very testing task. As the following table shows, 1.6 billion tonnes is over 1 per cent of the world's total biomass growth each year, and equivalent to a little less than a fifth of the biological material we produce on the world's croplands.

How much biomass is there?

Total amount of biomass on land	600 billion tonnes
Approximate gross rate of growth in biomass every year	115 billion tonnes
The biomass produced on agricultural land, e.g. food and straw	9 billion tonnes

Where would 1.6 billion tonnes of biomass come from? Part of it could be provided by agricultural wastes. Of the 9 billion tonnes of material produced on agricultural lands, probably less than 5 billion tonnes is actually food. The rest is straw, husks, leaves, and other currently unusable matter. We could use a substantial part of this waste for making char – though the benefit would be slightly reduced by the fact that some agricultural wastes are already ploughed back into the field, with some increase in productivity and carbon levels.

Furthermore, we will need to use the biomass from the world's lands to make the cellulosic ethanol discussed in Chapter 7 and for burning in the district heating plants referred to in Chapter 4. At first sight biochar seems to be one of a range of substantial extra demands that we will need to place on the world's agricultural, forest and savannah lands.

This may be too pessimistic an interpretation. Some of the biological material we need for biochar can come from the conversion of wooded areas containing slow-growing trees to what will be, in effect, charcoal factories using grasses, trees and shubs that develop much more rapidly. The world's surface has about 100 million square kilometres of vegetated land, including forests, savannah and cropland. At existing typical growth rates, to grow all the 1.6 billion tonnes of biomass we need might require us to convert up to 2 per cent of this area, or about 200 million hectares, to biochar production. This is an area four times the size of France, one of Europe's largest countries.

If we got our char half from forests and half from agricultural wastes, we would still need more than 1 per cent of all the world's green lands to become charcoal factories. This sounds an enormous task that is well beyond our capacity to organise. But if we doubled the typical weight produced by a hectare of forest by switching to faster-growing species, this becomes a much more manageable task. This idea has been greeted with horror by some green commentators who envisage biochar production becoming a repeat of the disaster of first-generation biofuels. The proponents of biochar need to address this point: if biochar becomes a valuable commodity, how do we stop local entrepreneurs and large companies cutting down forests in order to make it? I think there are three potential responses.

First, a sustained international policy of biochar manufacture will improve crop yields, particularly on degraded arable soils. So biochar will help the world feed its growing number of people and it may make good financial sense to use charcoal to improve soil fertility rather than burn it for fuel. Certainly Laurens Rademakers' 2009 results from Cameroon support the idea that biochar is so worthwhile for agricultural productivity that it will reduce the pressure on the world's forests rather than increasing it. If we can make more food on existing arable lands, we don't need to cut down forests.

Second, biochar will largely be made in poorer tropical nations where wage levels are low but biomass growth rates are typically high. It will be considerably less expensive to sequester carbon using tropical biochar than it would be in developed temperate countries. In fact, biochar could be a profoundly effective form of job creation in the poorest countries of the world.

Third, the processes of making biochar also produce useful heat, which can be used for domestic cooking if operated on a small scale. In larger biochar plants, the gases can be burnt to provide electricity through a simple turbine. It will be some years before we find out whether solar or wind technologies are better ways of supplying electricity in less prosperous parts of the world but a biochar kiln that burns wood gases in a turbine may be a cheap way of creating a reliable electricity supply in a remote country. It also has the advantage of being easily able to generate electricity at night or when the wind isn't blowing.

Some readers will still be aghast at any suggestion that we convert existing forests to plantations for biochar raw material. It is easy to understand the concern: anybody worried about global warming should be eager to maintain all the woodland we have today. But many types of tree are slow growing. If we replace these species with rapidly growing trees or grasses suitable for the region, we will eventually increase the uptake of carbon dioxide from the air. In a temperate climate with reasonable rainfall, a hectare of broad-leaved trees might increase in weight by 4 or 5 tonnes a year. Replacing these trees with fast-growing willow or miscanthus can triple or quadruple this weight gain, thereby massively increasing the potential for biochar carbon capture.

Of course, if we created a huge worldwide biochar industry, we would need to put in place stringent measures to protect biodiversity. In the drive to maximise biomass production for biochar and ethanol manufacture, we run the risk of creating areas of dangerous monocultures. One single type of tree might be planted on every spare hectare of land for hundreds of square kilometres. This would induce a catastrophic loss of diversity of animal and insect species in some parts of the world. It might also cause an increased possibility of disease and pest infestations that kill trees, or reduce their growth. So the world will need to control the planting of fast-growing trees to ensure sufficient variety of species. Once again, no-one can pretend that this will be easy task. We need economists to tell us how to create the incentive for landowners to grow as much biomass as possible while protecting soil and biodiversity and minimising the risk of widespread disease and infestation.

Some soils treated with biochar have another impressive characteristic not so far mentioned: they seem to give off far lower volumes of nitrous oxide. Nitrous oxide is a far worse global warming gas than carbon dioxide and, as we saw in the chapter on cellulosic ethanol, agricultural land is a prime source. In fact, for many crops grown in temperate regions, nitrous oxide emissions may be more important than the carbon dioxide produced by the growing and processing of the food.

This benefit is likely to be seen mainly in cooler countries and places where most artificial fertiliser is applied, though in one trial in tropical Colombia researchers found that adding biochar to grasslands reduced nitrous oxide emissions by 80 per cent. Output of methane, also a serious global warming gas, was cut almost to zero. Even if the addition of biochar is found to make little difference to agricultural productivity in soils that are already fertile, the impact on nitrous oxide emissions may be substantial. Although data on agricultural nitrous oxide emissions is poor, farming is becoming an important source of this greenhouse gas in some countries. Introducing biochar into the soil may be a cost effective way of reducing aggregate emissions.

Scientific knowledge of the best ways to use biochar is still developing. One key researcher in the field is Johannes Lehmann, a German-born academic based at Cornell University in upstate New

York. Lehmann and a former research student of his, Christoph Steiner, think that the most effective way to employ biochar to combat climate change is to work to replace traditional 'slash and burn' techniques of subsistence tropical agriculture with what the researchers call 'slash and char'. They point out that in many developing countries farmers move from area to area, cutting down and burning the forest as they go. The cleared land delivers reasonable yields for a couple of years but then rapidly declines in fertility and becomes unusable for a decade or more. If, instead, the farmer cut down the trees but then made biochar from them in a low-oxygen kiln, the charcoal could be added to the soil and keep it fertile for much longer.

On the best terra preta soils – which probably took several decades to create, many centuries ago – the fallow period is as little as six months after one crop has been harvested. Properly looked after, these soils never become unusable or have to be left for long periods of unproductive fallow. The slash and char technique could make a huge difference both to agricultural yields on weathered tropical soils and to the area of forest that has to be chopped down each year. The key task is to persuade millions of small-scale farmers across the tropics that slash and char is a better technique than their current methods. Once again, this may seem an intimidating task, requiring huge amounts of education in remote communities. But the impact on poverty and rural deprivation could repay the effort, even before measuring the global warming benefits.

Professor Lehmann is one of the great optimists of this story, offering a sense of the enormous potential of biochar. He says that by the end of this century, we could capture 9.5 billion tonnes of carbon each year simply by large-scale adoption of biochar manufacture in tropical agricultural systems. This is a striking figure: if we achieved this level of carbon capture today, atmospheric concentrations of carbon dioxide would be falling.

Kick-starting the biochar revolution

What do we need to do to get meaningful amounts of biochar into the world's soils? In the countries where we can install substantial numbers of large-scale plants, such as those that will be produced by BEST Energies or EPRIDA, all we probably need is for governments to acknowledge that biochar should be included in carbon trading schemes. During 2009, policymakers certainly became very interested in this option. A tonne of carbon stored in the soil has the same effect as capturing the same amount from a power station. So the operators of biochar plants need to be able to sell their carbon sequestration for the standard market rate.

At the autumn 2009 price of carbon dioxide in the European Commission's trading scheme, a tonne of carbon is worth almost $80. (A molecule of carbon dioxide weighs 3.67 times more than an atom of carbon.) In richer countries with good soils this might already be enough to make biochar commercially viable. Of course, the charcoal plants would additionally be able to sell their electricity and the biochar to farmers trying to improve soil productivity. If biochar also helps to reduce nitrous oxide emissions from the fertilisers added to the soil, the financial advantages would be even clearer. It may or may not make sense to build large-scale biochar plants in poorer countries. The capital costs of the equipment will be high and it may be cheaper to make biochar using large amounts of labour to construct charcoal ricks.

In less prosperous countries the fertility effect of adding biochar to the soil may be so substantial that it makes good sense even if the farmer is not remunerated for the storage of carbon. But in many parts of the world people live very close to the margins of survival. Rather than use the charcoal as a soil improver for next year's crop, households may decide to burn the charcoal instead, especially in areas short of firewood. Three things need to happen to make farmers in poor countries want to put charcoal on their fields. They need to be able to buy low-cost fuel-efficient stoves on financial terms they can comfortably manage. Second, particularly for those on the edge of survival, they need to be assured that biochar really does make a difference to soil fertility. Third, they also need to be brought into the world's carbon trading systems. It is not enough to reward just commercial biochar plants in rich countries; we also need to remunerate smaller peasant producers around the world.

The big problem, of course, is setting up a system that administers and monitors the scheme so that villagers in the tropics can get paid for sequestering carbon in the form of biochar, whether in tiny stoves or through the village's weekly charcoal-making session. No one denies that this is a substantial and perhaps even impossible challenge. But the rewards are potentially so great, both in climate change abatement and in poverty alleviation, that the global community needs to try to find a workable scheme. We will need an extensive monitoring organisation, able to reward those farmers maintaining forest cover as well as checking on improvements in soil carbon levels. Advances in satellite photography do make it much easier than even ten years ago to check that the amount of land maintained as forest is growing as a result of the biochar programme.

Speaking at a conference of biochar specialists in 2007, Australian Tim Flannery, one of the world's best respected earth scientists, fully endorsed their enthusiasm for using charcoal. 'Your technology offers the possibility of taking carbon dioxide out of the atmosphere ... and permanently sequestering that carbon in the soil,' he said. 'It does seem too good to be true, but I've looked at it from every angle, and I fail to see the fault in the system.'

It would be very easy to dismiss biochar by saying that it requires too many new manufacturing plants, or that the carbon storage in the soil is too difficult to monitor. But unlike most of the other carbon reduction ideas in this book, biochar involves no great technological uncertainties or unknown costs. The problems are essentially managerial. And free markets are good at solving managerial problems. Biochar will need to be substantially aided over the next decade, but once established under the aegis of a reliable carbon price, it will support itself. The fact that biochar is also disproportionately of benefit to the poorer and rural parts of the developing world should make us even more enthusiastic.

Soil and forests

Improving the planet's carbon sinks

We tend to portray the climate change problem as one-dimensional: the simple result of burning fossil fuels. This is wrong. The issues are actually vastly more complicated and, perhaps paradoxically, this gives us greater reason for hope. In particular, the continuous circulation of carbon between the atmosphere, the oceans, soils and plants provides us with the wherewithal to counteract the impact of fossil fuel emissions. Even if we can't quickly reverse the growth in fossil fuel use, we have some tools to offset these emissions by increasing the take-up of carbon elsewhere in the ecosystem. We can use carbon sinks to 'swallow' some of the excess carbon dioxide arising from the burning of fuels.

In the last chapter, we examined one easy way of getting carbon dioxide out of the atmosphere and safely stored as carbon in the soil. This chapter also focuses on soil and, secondarily, on plants. The reason for this attention is that the world's soils do appear to be the best available repository for excess carbon resulting from mankind's action. As the chapter on biochar showed, much of the world's soil contains inadequate stores of carbon and increases will be beneficial to vegetation growth, which is also an important way of storing extra carbon. If we can make more food on existing arable lands, we don't need to cut down forests.

In contrast, we will find it difficult to use the oceans to increase their carbon storage. If we injected carbon dioxide into surface waters, the impact on the ocean's acidity would be severe, and anyway the carbon dioxide would almost certainly transfer back to the atmosphere quite quickly. There is already some worrying evidence that the capacity of the seas to hold the existing stock of carbon dioxide is reaching its limit. Injecting carbon dioxide into the very deep ocean where it will liquefy probably does not entail these problems but is technically difficult and likely to be very expensive.

So we need to do everything we can to improve the world's carbon sinks and lock as much carbon dioxide as possible into soils, trees and plants. Adding charcoal to the soil works best for arable lands in tropical lands and areas of degraded soil. The biofixation techniques examined in this final chapter will improve the carbon content of pastoral lands used for grazing animals and also add to the productivity of temperate croplands.

Finally, I look at the best and cheapest ways of reducing the rate of deforestation in the world. As forests are cut down and burnt, they give up their store of carbon to the air. Large acreages of woodland disappear every year, increasing greenhouse gas emissions as well as creating other severe ecological problems, such as desertification, soil erosion, flash floods in wet seasons, and drought in the drier portions of the years. Arresting the loss of trees may be the cheapest of all techniques for carbon capture and will also improve the long-term ability of the world to feed itself and support its rural populations.

Scientists have a rule of thumb which states that about half of the world's emissions stay in the atmosphere once put there by man's activities. There's some disagreement about how much of the rest is absorbed by the oceans and how much by forests, soils and plants, as well as by the weathering of rocks, a process that absorbs carbon dioxide. Probably more goes into the oceans than into land 'sinks' but we cannot be completely sure. Some people think it varies greatly from year to year.

But we can be reasonably certain that only about half of the carbon dioxide from fossil fuels and other sources ends up in the atmosphere. We can estimate both the weight of the earth's atmosphere and the weight of new carbon dioxide emitted every year. If all the carbon dioxide that the world produced ended up in the air, the atmospheric concentrations of carbon dioxide would rise by about 4 ppm every year, not the 2 ppm we estimated in the chapter on biochar. However, the evidence is increasingly strong that, little by little, the world's seas and lands are becoming less effective at capturing our greenhouse gases. A slightly larger fraction of what we emit today is ending up in the atmosphere, rather than being safely sequestered. There's a tone of rising alarm about this subject in the major scientific journals.

The decline in the ability of the oceans and the land to capture our emissions is likely to continue, and possibly accelerate unless we take measures soon. For example, we know that the ability of the seas to take in carbon dioxide is affected by the water temperature. Higher temperatures cause a decline in the ocean's ability to dissolve carbon dioxide. Unsurprisingly, the temperature of the top layer of the world's water masses is generally getting hotter. So, all other things being equal, the amount of new carbon dioxide stored in the world's seas and lakes is likely to fall.

This chapter examines the most effective and simplest ways in which we can retain and enhance the land-based stores of carbon – soils and forests – which would otherwise add to the inventory in the earth's atmosphere. Do these qualify as a 'technology' in the words of the title of this book? We are accustomed to applying this word to advances in electronics or perhaps in medicine, but it actually has a much wider meaning. One of the Oxford English Dictionary's definitions is 'a particular practical or industrial art' and it is under this rubric that I include the world's efforts to find ways to productively store carbon in soils and forests.

Getting the most from the land

One thought crops up at multiple points in this book: we will not be able to deal with global warming without addressing the issue of how we use the world's land. Land provides our food and, in most of the world, our cooking fuel, in the form of wood, vegetation and dung. In a future low-carbon world, we will also be asking our soils to grow plant matter for cellulosic biofuels, district heat and power plants and biochar. The urgent need to increase food production to feed an ever-growing population appears to be squarely in conflict with these new demands. We have seen the first, and very frightening, illustration of this conflict in the impact of first-generation biofuels on food prices.

So the world faces an almighty challenge. We need to increase the productivity of the world's lands, so that they can produce the food for the 9 billion people likely to be sharing the planet in a few decades' time. At the same time, we need to use our soils to produce far more woody matter for fuel, heat, electricity and biochar. Equally importantly, we must also improve the ability of the world's soils to retain carbon, and ideally to soak up more. These objectives appear to be irreconcilable. For example, more intense exploitation of the world's soils in an effort to produce more grain will probably decrease the amount of carbon that the land can store. If we devote a larger fraction of the earth's surface to growing trees or energy crops, the space available for crops will decline. How do we resolve these apparently incompatible objectives?

As I've already said, one obvious answer might be for the world to eat less intensively farmed meat. About 35 per cent of the world's cereal grains get fed to animals. Even if nothing else changed, we could probably cope with a world population of 9 billion if none of us ate beef, lamb or pork. Unfortunately, as the luckier half of the world population gets more prosperous, we tend to consume more meat, increasing the amount of grain eaten by animals and therefore reducing the supply available to those who rely on cereal grains as their primary source of calories. For most people, increasing consumption of animal protein is an important benefit arising from higher levels of prosperity. For that reason, it will prove very difficult to cut the link between increasing wealth and industrial meat production, so I don't propose this as a solution to the conflict between our food needs and the requirements for low-carbon energy sources.

In fact, later in this chapter I suggest that better grazing practices may allow greater amounts of animal husbandry in many areas of the world. Although herbivores will always be significant producers of methane (created as a by-product of the digestion of cellulose), the little research on this topic suggests that animals kept outdoors and eating only grasses produce far less of this greenhouse gas than factory-farmed cattle. They also do not eat cereal grains and remove them from the food supply. In reality, then, it is not meat that is the problem: it is industrially farmed meat.

Instead of simply concentrating on meat production, the focus should be on maintaining and improving the level of carbon in the world's soils. As the previous chapter showed, the world's soils contain about twice as much carbon as is held in the air. So small changes in the composition of the soil can result in significant flows of carbon into or out of the atmosphere. Agricultural practices around the world must be geared in part to maintaining and improving the ability of soils, whether cropland, pasture or forest, to hold carbon.

Zero-till

Intensive cereal farming tends to reduce the amount of carbon in the soil through a variety of mechanisms. Ploughing the soil exposes organic matter such as humus to the oxygen in the air, and makes it likely to rot. Exposing the bare soil to rainfall after harvest increases the erosion of the normally carbon-rich top layer. Removal of straw and other waste matter similarly cuts the amount of carbon left in the soil. These outcomes are bad for the farmer, because the complex carbon-based molecules in soil humus are invaluable for maintaining the structure of the soil and for making nutrients available to plants. They are also bad for the planetary atmosphere because the carbon in the soil degrades into simpler molecules such as carbon dioxide and methane which then add to the stock of greenhouse gases in the atmosphere. All things being equal, soils with high levels of complex carbon molecules will be good for the climate and good for agricultural productivity.

In tropical soils used for growing crops, biochar will help improve soil carbon levels. On temperate cereal-producing soils, however, the most effective way of increasing soil carbon levels may be to shift to a form of agriculture known as 'zero-till'. Farms using this approach never use a plough to turn over the soil. The proponents of the zero-till technique believe that ploughing is counterproductive because it reduces the soil carbon content and increases the loss of valuable moisture. Perhaps surprisingly to those of us who love the smell and appearance of a freshly turned cereal field, ploughing has little direct benefit to the crop. It reduces weed growth and helps expose some harmful insects to birds and other predators. But these benefits are offset by the potentially detrimental effects on the soil.

Zero-till farmers disturb the soil as little as possible. They plant the seeds in a row, accompanied by fertiliser. Once the main crop is harvested, a second crop is planted to provide a cover for the soil to prevent erosion and any loss of fertility. Often called 'green manure', this cover crop is cut down and left on the surface of the soil to slowly rot and fertilise the soil. These dead stalks and leaves help protect the soil from erosion caused by heavy rains and provide a rich source of organic material that earthworms can use to improve the quality of the lower soil. The following year, the seeds are planted through this material, known to farmers as 'trash'. Regular varying of the crop prevents the build up of destructive pests and diseases.

The evidence that zero-till farming increases soil carbon levels, partly by minimising erosion by water, is now compelling. It may not work on every type of soil, and results tend to vary from year to year, but switching from conventional arable farming techniques to zero-till seems to reverse the trend for heavily cropped soils to lose carbon. In addition, crop yields seem to improve, although for many farmers there is an initial period during which productivity suffers. In other words, though the debate is not yet over, zero-till could increase the tonnage of grain the world harvests every year *and* turn intensively farmed soils from being net emitters of carbon to being an increasingly valuable store.

Of course, the zero-till approach has its drawbacks, at least in some eyes. It uses large amounts of herbicides to control weeds that would otherwise have been averted by deep ploughing. In fact, zero-till may work best with crops that have been genetically modified to withstand the most common weedkillers so that herbicide can be applied while the crop is growing. Either way, the zero-till approach is largely incompatible with organic farming, which bans the use of genetically modified seeds and herbicides. Organic farming, for all its potential benefits, uses regular ploughing to ensure that weeds do not become dominant, thus tending to reduce the ability of the soil to store carbon.

From a global warming point of view, organic farming has another disadvantage. Many organic farms produce cereal yields of little more than half the level of similar farms farmed in a conventional way. The difference largely arises because organic farming builds up fertility by planting nitrogen fixing plants such as clovers for several years after each grain crop rather than applying artificial fertilisers. Put another way, if the entire world's cropland was farmed organic- ally, it might take twice as much land to produce the food to feed the global population. Organic farming systems usually also require a large number of animals to provide the manure that fertilises the cereal fields. In many countries, these animals need substantial amounts of winter feed. These aspects of organic farming all assist in increasing the pressure on the world's limited stock of high quality farmland, increasing the incentives to cut down forests.

Organic farming also has substantial advantages. For example, it assists in the reduction of nitrous oxide emissions because it avoids the use of artificial fertilisers. And the long crop rotations probably help restore the soil carbon losses caused by frequent ploughing. The point I want to make is not that organic farming is bad, but rather that 'zero-till' techniques may be better at simultaneously providing high yields of foodstuffs and maintaining soil carbon levels.

As a result of the large advantages it offers on most soils, a large percentage of total arable land in some parts of the world has now been converted to zero-till cultivation. Over 20 per cent of Ameri- can farmlands are now avoiding the plough, and the figure is even higher in Brazil and Canada. Zero-till has made slower progress outside the Americas and Australia, perhaps because in Europe and other high-latitude zones low levels of soil carbon are not yet a threat to production levels. Soil carbon losses in conventionally tilled soils are rapid only above temperatures of about 25°C, a threshold that is reached only for a few days each year in high latitudes. Nevertheless, there is increasing evidence that even in cooler countries, zero-till will significantly help maintain or increase the amount of carbon stored in arable soils.

In Brazil, earlier phases of intensive cultivation may have cost 30 or 50 per cent of the original soil carbon in the dry 'cerrado' region south of Amazonia. Although research produces very varied results, a typical hectare of soil in this area, if cultivated with zero- till techniques, absorbs 0.6 to 0.7 tonnes of new carbon each year.

This equates to well over 2 tonnes of carbon dioxide, suggesting that if the grain producing area of the entire world (about 700 million hectares) were switched to zero-till, the sequestration might be worth almost 2 billion tonnes of carbon dioxide a year, between 5 and 10 per cent of today's total emissions. Perhaps the impact in higher latitudes will be less than it is in Brazil – we really don't have good information on this yet – but a robust global climate change policy would probably be incentivising the introduction of zero-till techniques. In addition, a switch to zero-till techniques will reduce the amount of fossil fuel used on farms because of more limited use of tractors. The extra use of weedkiller will only partly wipe away this gain.

Much as we saw with biochar, a further additional benefit of zero-till may be the reduction in nitrous oxide emissions from soil. On a zero-till farm, smaller amounts of fertiliser may be applied, sometimes placed next to the seed in the sowing process. Researchers are not yet completely confident, but it seems that the limited and highly targeted application of nitrogen fertilisers in a zero-till system may potentially provide another significant advantage over conventional farming methods, which spread the fertiliser over the whole surface of the field during the growing season.

Along with the other techniques described in this chapter, a huge increase in the use of zero-till farming may be an excellent way of keeping carbon in intensively farmed soils. In the long run, however, there may be an even more effective way of improving the world's carbon sinks: reviving degraded grasslands.

Reviving grasslands

Australian Tony Lovell isn't a farmer, but his attempts to demonstrate the importance of increasing the carbon levels of soils have begun to attract attention. He persuasively argues that the easiest and best way to reduce the excess carbon dioxide in the earth's atmosphere is to improve the soil health of dry pastoral lands Overgrazing and erosion over the decades have taken much of the carbon out of most of the world's enormous areas of ranchland.

Tony is an accountant by training and within a few minutes of starting a conversation with him, you get drawn into doing some

*A Mexican ranch with almost unusable land and the
next-door ranch with abundant vegetation*

ecological arithmetic. But first he shows you some photographs, starting with a ranch in Mexico: empty, bare soils, with very little vegetation. Next is a photo of the next-door property, a few miles away. Same soils, same rainfall, but abundant growth and a healthy ecosystem supporting cattle and a range of wildlife. The difference is astounding. What causes the huge variation? According to Lovell, it's simply a matter of how animals graze the land.

Before people came along, Lovell explains, huge herds of herbivores such as bison or wildebeest moved across the plains of the Americas and Africa. The animals stayed together in a group because of their fear of predators such as lions. They kept moving because otherwise the grasses would be insufficient for the large group. Once a stretch of grassland had been eaten, the herd would not return to the areas for several weeks, by which time the crop would have grown back.

Mankind changed all this. The predators of the herbivores were gradually killed, allowing the animals to roam singly or in small groups. Liberated from dangerous carnivores, these animals could repeatedly feed on the same areas, keeping plant growth to a minimum. There were no long recovery periods for the grasses.

Lovell points out the importance of this. Unlike a tree, he says, most of the weight of a grass lies under the ground. Whereas trees keep their stored carbon in their trunks and branches, the carbon captured by grasses ends up mainly in their large root systems. The roots of some grasses can go metres into the ground. But when the grass is eaten, and its height reduced, the plant sloughs off some of its root system. Within minutes, Lovell says, a grass that has had much of its leaf system removed by a grazing animal will start to reduce the length of its roots. This previously stored carbon gradually returns to the atmosphere as the dead matter rots away. If the grass is then not eaten for a period, the roots will be gradually re-established. But grasses that are repeatedly cropped will never again build up long carbon-rich roots, cutting the amount of carbon in the soil. Partly as a result of this, the ground will hold less water, and almost all vegetation will eventually die. The vicious circle is completed by enhanced rates of erosion arising from the lack of roots to bind the soil together and the absence of vegetation to protect the surface. Occasional flash floods remove most of the carbon that remains.

Pastoral soils that have lost most of their carbon can be seen all over the world, but particularly where the rainfall is highly seasonal. Lovell calls these areas 'brittle', an expression first employed by his mentor, the great Zimbabwean pastoralist Allan Savory, who now runs a centre in the US seeking to improve the management of the world's vast rangelands. A brittle area has long periods, of perhaps six months or more, in which virtually no rain falls. In non-brittle areas, such as the croplands of northern Europe, misuse of the land does not result in a desertification: the regular rain allows the land to partly recover. But in brittle regions, the damage caused by overgrazing will usually tip the land into what is generally thought to be an irreversible decline.

Poor grazing practices – sometimes imposed a hundred or so years ago by new immigrant farmers from water-rich regions – helped reduce the carbon in the soil. These practices also damage food production, by massively reducing the ability of the land to feed animals. The impact on poor rural communities can be devastating. For example, in recent years the threat from ever-growing deserts in China has forced the government to ban the grazing of animals over vast tracts of land in the west of the country. By completely stopping any use of the land by pastoral animals, the Chinese are trying to help restore the health of the soil, increasing its ability to hold water and grow grasses with deeper roots. This may well have been a necessary step but it has adversely changed the lives of many thousands of pastoralists. Unless stopped by draconian measures such as these, overgrazing will eventually help create deserts in many of the driest parts of the world.

According to Lovell, Savory and others, reversing the process that is degrading huge expanses of land around the world is difficult but manageable. The task is to recreate the way herbivores used to crop the land – in large groups that moved swiftly across the landscape, not returning until the grass has recovered. So instead of allowing animals to range in small groups across the entire landscape, the herbivores need to be herded carefully. A sheep farmer might move the livestock every four or five days from enclosed pens, not coming back to the same place for ten weeks. Or cattle could be moved by herdsmen opening and closing water points and actively herding the animals from one place to another.

Lovell points out that it is not the fact that the land is grazed that causes the problems. Through their dung and urine, the animals return almost all the carbon and nutrients taken from the soil and, paradoxically, by trampling down plants, the beasts help to maintain a layer of decomposing organic matter which protects the soil. As Lovell and others are at pains to point out, proper grazing management can actually substantially *increase* the number of animals that a ranch can support. An 'overgrazed' landscape is not one that has had too many animals on it, but one on which grazing is not properly controlled. By contrast, good management means water is retained and grasses grow quickly. Patches of animal manure provide places for new seeds to safely germinate and help build the soil.

Some scientists have known for generations that controlling the movement of pastoral animals is crucial to maintaining soil health in brittle lands. One of the classic books about the human influence on climate put this very clearly in 1977. In *Climate of Hunger*, Reid Bryson and Thomas Murray wrote that:

> A fence built round a large field in a desert brought forth abundant wild grasses in two years, without planting or irrigation, mostly by keeping out the men and the goats. As similar experiments in the western US show, grasses do grow if grazing pressure is lifted.

In fact, properly grazed land, fertilised by manure, will often show faster vegetation growth than completely abandoned land. One of Lovell's most extraordinary photographs shows an enormous 450 hectare pile of copper mine tailings in Arizona. After sixty years, the massive hill had virtually no vegetation cover. Eroded gullies ran down the side. Then cattle were introduced on part of the pile. Fed with hay, the animals' manure gradually provided an environment in which plants could establish themselves. The animals' feet pushed remnants of their feed, covered in fertile dung and urine, into some of the holes in the surface, creating good environments for grass growth. Eventually the cattle created a soil 30 centimetres thick. The soil substance is, of course, almost entirely organic matter that has grown by sequestering atmospheric carbon dioxide through the photosynthesis process. The roots of the grasses have also bound the surface of the mound together, meaning that erosion has stopped and the carbon will be permanently stored, provided the careful grazing regime continues.

Another photograph shows an adjacent area that was kept free from animals and instead treated with grass seeds and regularly watered. This attempt to stabilise the mound and build a resilient soil completely failed. In the picture, the surface is still completely bare and scarred by deep gullies.

Photographs like these make an utterly compelling case. So why hasn't the world rapidly adopted the advice of Allan Savory and his followers? The answer to this question is partly that any form of farming innovation is highly risky. If things go wrong, the farmer could be ruined. In addition, many farmers already live on the edge of economic failure (partly as a result of falling income caused by bad grazing practices). In these circumstances, the tendency to resist new ideas is understandable. But once a few farms move to carbon-enhancing grazing management, many of their neighbours eventually follow. Of course, there may be other problems as well. Just as I finished this chapter, I met a distinguished Namibian architect whose father had followed Savory's principles in managing the family farm in that dry and hostile climate in southern Africa. She agreed that soil management was vital and had helped her father grow cereal crops in an area where water shortages and erosion made this very unusual. But in the case of her father's farm, there had been a compensating disadvantage. One year a severe fire had overrun part of the land. The farm's sheep had all been fenced into a small area, in line with Savory's recommendations, and were unable to escape the blaze. Grazing practices that allowed the sheep to roam freely would have avoided this severe blow to the farm's fortunes.

Despite problems such as these, better management of grazing will generally be good for farm incomes, and also for biodiversity and soil fertility. What about the impact on global warming? Tony Lovell's figures suggest a remarkable answer. Improving grazing lands may be the single most important step we can take to combat climate change and to feed the world. Lovell says that the world has about 5 billion hectares of 'brittle' pastureland, perhaps seven times more than the land devoted to crops, almost all of which could be managed for improved food production and enhanced carbon storage.

Let's look at what happened when one thoughtful farmer switched to better grazing management. Lovell quotes figures from a pioneer in the dry Gulgong area of the New South Wales tablelands whose 900 hectares grows some wheat but also provide the grazing land for over 4,000 sheep. This innovator increased the carbon in his farm's soil by over 50 per cent over a ten-year period by careful improvement of his grazing practices and by planting his wheat in rows interspersed with pasture land. The average level of soil carbon rose from under 2 per cent to over 3 per cent of the mass of the soil.

This sounds a small difference. It is not. Making this apparently insignificant improvement to the top 30 centimetres of the soil sequesters a massive 100 tonnes of carbon dioxide per hectare. It also substantially improves the productivity of the land, and makes it more resistant to the impact of drought, an extremely important consideration in Australia, where climate change is likely to cause long-term water shortages over much of the continent. In fact, most of the currently dry areas of the world are likely to be at substantially enhanced risk of catastrophic drought in our hotter world.

The experience of this pioneer and many others is that after a few years of transition the profitability of farming is improved, just as with zero-till cultivation. In other words, this is a very low-cost form of carbon sequestration. Like biochar and growing algae, it may be a much cheaper form of carbon capture than the techniques for capturing emissions from power stations described in Chapter 8. It also has the profound advantage of being ready to roll out, unlike carbon capture technologies that are still being developed.

There are about 200 million hectares of brittle grazing land in Australia alone. So if we could find a way of adding 1 per cent to the carbon content of all these soils we would be storing almost 20 billion tonnes of carbon dioxide, about 80 per cent of the world's current annual emissions. Similar dry ranchlands can be found across the tropical and subtropical world, including much of sub-Saharan Africa and large areas of South America, as well as the south-western US and parts of Mexico. It would be foolish to suggest that a single management approach will work in all these regions. What succeeds in countries with mobile herdsmen, such as in some parts of Africa, Asia and even southern Europe, will not succeed in more settled pastoral communities, or in areas that are almost desert. The amount of extra carbon that can be stored will also vary. But the potential is huge: if 1 per cent extra can be achieved across just half the world's brittle lands, that would soak up the world's current emissions for over ten years.

There is a very strong argument for including soil sequestration in the European carbon market, and any future US or global systems. The carbon dioxide moved from the atmosphere to the soil as a result of better land management has a value to the global community equivalent to a reduction in power station emissions. A farmer storing an extra 100 tonnes of carbon dioxide in a hectare of land should be able to receive a reward of $2,000 for his or her valuable effort. For farmers struggling to survive on increasingly degraded land and facing the prospect of more erratic rainfall as a result of climate change, this amount of money is a vital addition to farm income. A 1,000-hectare ranch, perhaps on the Argentinian pampas or in the south-west US, could earn over $3m by increasing soil carbon levels by 1 per cent over a ten-year period.

Of course we can expect abuses of a system like this, and it would be troublesome and complex to administer, but not necessarily more so than the Clean Development Mechanism (CDM), one of the Kyoto Protocol's carbon trading tools. Many projects that have qualified for payments under the CDM have turned out to be little more than hoaxes, but soil sequestration schemes are less likely to be misused. After all, storing carbon this way is cheap, technologically simple, and easy to validate. Success might mean eventually returning the amount of organic matter in the soil to the levels prior to mankind's intervention. The benefits of this would be much wider than just climate change, with positive implications for food supply, water availability, and the incomes of struggling farmers.

Scientists have advanced many different high-technology

solutions for dealing with the existing stock of CO_2 in the atmosphere. As we'll see in the epilogue of this book, many of these ideas are expensive or potentially dangerous or both, and the results are uncertain. By contrast, Tony Lovell's vision is simple and virtually risk-free: to return the earth's pasture lands to their condition prior to the not entirely benevolent takeover by mankind. The side-effects of improved grazing management, such as improved biodiversity and productivity, are likely to be benign.

Along with zero-till farming, better grazing management shows that there needn't be any conflict between our climate change and agricultural objectives.

Forests

Many of us have a strong sense that the world is rapidly losing its forest cover. In particular, we all understand the threats to the Amazonian rainforest, under continuous pressure from illegal logging and conversion to soy plantations and cattle ranches. Similarly, Indonesia is switching virgin forest to huge areas of palm oil plantations. Deforestation in these countries and elsewhere is increasing the pace of global warming by adding carbon to the air.

Although forest loss is a hugely significant issue that has captured media attention, the reality is possibly slightly less bad than most people imagine. Although the world is losing its forest cover, the rate of loss is probably now declining. The UN Environment Programme (UNEP) says that global deforestation amounts to about 13 million hectares a year, out of a total of about 4 billion hectares of woodland, or somewhat less than 0.5 per cent a year. This reduction is partly compensated by the creation of new forests, although no one should pretend that establishing new plantations compensates for the many detrimental effects of the destruction of virgin woodland.

That point aside, we can estimate that the net loss of forested area is probably between 7 and 8 million hectares a year, or 0.2 per cent of the world's wooded area. Forests are actually increasing in size in Europe and in some other parts of the world. Largely to try to prevent further desertification of large parts of its interior regions, China has planted over 50 billion trees in the last two decades or so, creating 54 million hectares of new forest. This has helped slow the march of the deserts that cover over a quarter of the land area of this huge country. The success of the Chinese investment in trees proves that deforestation is not an inevitable process which the world can do nothing about. Well-organised countries can improve the level of tree cover. Nevertheless, deforestation is still playing an important part in increasing the levels of greenhouse gases in the atmosphere and requires an urgent response from us.

Wood is approximately 50 per cent carbon. When forest is lost, this carbon is largely transferred to the atmosphere, particularly if the wood is burnt. UNEP says that about 4 gigatonnes of carbon dioxide, containing over a billion tonnes of carbon, are lost from forest biomass every year. This equates to perhaps 15 per cent of the total additions of carbon dioxide to the atmosphere. This is a huge figure and scientists are right to repeatedly draw attention to it. But the optimists among us need to remind ourselves that new carefully managed wood plantations in the humid tropics could completely replace this annual loss by the natural yearly growth of an extra 80 or 100 million hectares. This would involve adding only 2 per cent or so to the world's forested area or less than twice the amount the Chinese have already created. Forest plantations may not be to everybody's liking, not least because they tend to use a small number of species and introduce limited biodiversity of animal and plant life. But a well-managed planting programme in the tropics will foster trees that grow ten or twenty times as fast as unmanaged woodland in the temperate zone.

By intelligently planting the right tree species in woodland plantations in places where growth is likely to be rapid, we can roll back some of the appalling effects of forest loss in the Amazon and elsewhere. This point must come with a caveat. The gradual loss of the Amazonian rainforest will eventually not just affect greenhouse gas levels but will also change the world's weather patterns, probably making large areas in the western hemisphere much drier than at present. By stressing that an active programme of reforestation is a vital ingredient in combating climate change, I don't want to give an impression that I think we can simply let the Amazonian and Indonesian logging continue.

Deforestation in these areas looks an intractable problem. Some forest clearing is occurring as a result of large commercial farmers wanting to add to their estates. But it also happens partly as a result of the decisions of millions of people trying to create new land on which to grow food. In many countries land tenure law does not give clear ownership rights over these trees, making the protection of forest very difficult. States such as Brazil may be able to restrict the loss of land to soy crops and beef ranching, but telling the landless poor that they should not try to feed themselves by growing crops on cleared land is a policy that is unlikely to succeed anywhere in the world.

However, the longer-run effect of widespread woodland loss is to increase soil erosion and cut fertility. It will also probably change local microclimates, eventually reducing total agricultural production. The chapter on biochar explained how tropical 'slash and burn' techniques could be amended to keep some proportion of newly cleared wood as carbon in the soil, helping to improve fertility and reduce erosion. This will help reduce soil carbon losses.

But there are other things we can do as well. Probably the single most important change we could make would be to reduce the amount of wood cut down for cooking fuel. Energy-efficient biochar stoves would help but many other types of improved stove design have been trialled throughout the world. Some are made from ceramic or brick components and others from metal. By capturing more of the heat from burning wood, they reduce the amount of firewood needed, and therefore cut the amount of local forest loss. Particularly in Africa, where UNEP says that half the net loss of forest is caused by the need for cooking fuel, efficient stoves could significantly reduce the rate of deforestation. As we've seen, they also burn more cleanly, avoiding dangerous local air pollution.

Of course, what looks simple and economical on paper does not always work quite as well as expected in the field. Studies looking at the roll out of new stoves have shown that although the savings in wood fuel use might be as much as 50 per cent, this target is often not achieved. People don't always operate them in the most effective way – or even abandon them because they take longer to cook the food. Hungry people don't want to wait any longer than necessary to eat.

Nevertheless, a worldwide effort to improve the efficiency of cooking remains crucial to addressing the forest loss problem. At the moment, the drive to reduce the amount of wood used in stoves is being led by a mixed bag of small charities and 'carbon offset' specialists (organisations which establish carbon reduction projects to counterbalance emissions caused by their customers' air travel and other activities). One example is UK-based Climate Care, which spends much of its income developing and marketing new types of stoves in countries as different as Honduras and Bangladesh. This work is a very useful start, but the world's richer countries should contemplate a much larger and more widespread effort to reduce the number of trees cut down for cooking. About 2 billion people (nearly a third of the world's population) cook using wood or dung. We cannot really be sure of these figures, but some estimates suggest that each person needs about 800 kilograms of cooking fuel each year. Multiply the two numbers together and we get a global figure of about 1.6 gigatonnes of wood and dung used each year, of which about 800 million tonnes (50 per cent) is carbon. UNEP says that forest loss produces about 1.1 gigatonnes of carbon a year, meaning that cooking accounts for about three-quarters of the net loss of forest volume. Put another way, if we could cut the amount of wood fuel used in cooking stoves in half, this alone would reduce net man-made additions of carbon dioxide to the atmosphere by over 5 per cent.

Fuel-efficient stoves are one of many ways of reducing wood fuel use in tropical countries. Other agencies are working on solar cookers, including various devices that focus the sun's rays on to a container of water. These solar cookers use exactly the same principle as employed in the concentrated solar power plants described in Chapter 2: reflecting sunlight on to a small area using mirrors and generating intense heat in a small area.

Another promising technology is the use of biogas collectors. When human, animal and plant wastes rot down, they give off methane, particularly if kept in enclosed containers or left in a rubbish dump or landfill site. In most parts of the world, this methane vents to the open air but if it is collected it can be burnt in simple stoves, replacing the need for woody fuels. (What we call natural gas is almost pure methane.)

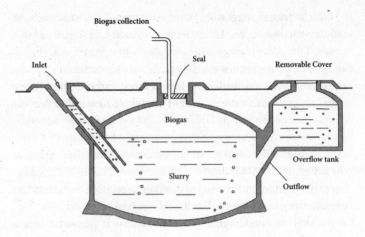

Drawing of a domestic biogas collector

Alongside its policy of large-scale reforestation, the Chinese government has heavily promoted the biogas stoves to reduce the need for families to burn wood. In rural areas, concrete or brick-lined pits next to houses collect latrine waste and other materials that will rot, such as straw from fields. The pit is covered and sometimes a greenhouse is put on top. The greenhouse is warmed by the heat from decomposition and, in turn, the greenhouse helps insulate the pit so that temperatures can be high enough for the rotting process to continue. The resulting methane is taken by pipe to a cooking stove and sometimes to gas lamps. When the decomposition has finished, the waste matter, now sterile, can be applied as fertiliser to fields or even fed to pigs. In the past, human wastes might have been put directly on to fields, from where pathogens could flow into local watercourses, so there are also substantial health benefits to the new approach.

A biogas cooking stove may not completely remove the need for other sources of energy for cooking. In times of low ambient temperature, for example, the rotting process will slow down and the amount of methane produced will tail off. At such times the house may still need wood fuels for cooking. However, in most of rural China, average temperatures are high enough to produce enough methane for at least six or eight months a year.

Official figures suggest that over 18 million Chinese households now have biogas stoves. Estimates of the amount of firewood saved can be little more than guesses, but the figure could be as high as 30 or 40 million tonnes every year. This is equivalent to up to 2 per cent of the total worldwide loss of forest carbon. The Chinese government says that the number of household biogas digesters can be increased eightfold, implying further huge potential reductions in the use of wood fuels as well as improved soil fertility and lower rates of waterborne disease. As with the tree-planting programme, the campaign to install digesters to decompose waste in China has been driven more by the need to avoid deforestation than to reduce greenhouse gas emissions. But the two go hand in hand.

As well as encouraging the installation of domestic biogas systems, the government has sponsored the setting up of waste digestion systems in animal rearing units. These larger scale anaerobic digesters then pipe methane – again for cooking and lighting – to local villages. Previously burnt or simply thrown away, agricultural straws and stalks are also increasingly digested in these methane-producing tanks. Copied across the world, these cheap, simple and benign technologies can systematically reduce the amount of wood needed for cooking.

My aim in the last few pages is not to suggest that deforestation is an uncomplicated issue to address. It is not. The pressure in some parts of the world to cut down trees for fuel or to clear agricultural land is intense. Eventually, as the Bali climate change conference decided in 2007, the world will have to pay local people to maintain forest lands, particularly in areas where population pressure is at its most intense. But until we have put in place a system of appropriate financial incentives, the right thing to do may be for the rich world to focus on the three simple technologies just discussed: fuel-efficient stoves, solar cookers and, perhaps most importantly, biogas tanks.

Each of these measures reduces the need to cut down trees, and biogas digesters, by producing fertiliser, may also increase production from arable lands. As with biochar, zero-till and improved grazing management, it's a way of saving carbon while simultaneously boosting food production. These techniques are the simplest and most basic of the portfolio of solutions to climate change, but they may also end up by being the most effective.

As a footnote, it's interesting to note that, whereas China has over 3,500 biogas plants treating animal wastes on commercial farms (in addition to the 18 million domestic units), the UK has only about 20. The countries that talk a lot about climate change are not always the ones doing the most about it.

Putting it all together

Are the ten technologies enough to save the planet?

Policymakers around the world believe that in order to keep the global temperature rise below 2°C we must ensure that atmospheric concentrations of greenhouse gases eventually stabilise at no more than 450 ppm. This includes CO_2 and other gases expressed in equivalent terms. Broadly speaking, this means that the world's industrialised nations will need to reduce their emissions to 20 per cent of today's levels by 2050. Some people say that the reduction will have to be even greater.

The rate at which this cut is achieved can vary. We might choose to wait until 2020 before reducing our emissions. But leaving it this late would mean extremely fast annual decreases were then necessary. The global consensus is that the appropriate way to get emissions down to 20 per cent of the current level is to halt the rise in annual emissions soon, and then pursue a less rapid reduction plan. But this is not going to be easy. Even if we ensure that the emissions of the industrial world hit their maximum in 2010, we will need to cut emissions by 4 per cent a year thereafter if we are to reach the 2050 target.

If we make substantial progress in each of the ten opportunities described in this book, will we be able to reduce fossil fuel use in advanced economies fast enough? And can we counterbalance the emissions of methane and other greenhouse gases by improving carbon storage in soils and in plant matter? The best way to address these questions is to examine how energy is used in advanced economies and then estimate how much can be switched from fossil fuels to green sources. If progress is good, where might we able to get to by 2025?

Let's look first at how much fossil fuel energy the modern world uses. We can calculate the total energy use of a country by adding up the volume of fuels used, multiplying by their energy content, and then working out a figure for each inhabitant. The numbers are large. In a typical European country, around 50,000 kilowatt-hours of energy are used to support the lifestyle of each person. The figure is approximately twice as high in North America, and is lower in Japan. The fast-industrialising countries of Asia and elsewhere are lower still, at perhaps a quarter of the European level.

To express this figure another way, 50,000 kilowatt-hours per year is equivalent to a continuous stream of 5,000 or 6,000 watts – comparable to two electric kettles boiling day and night.

Some of this energy use is obvious to us – for example the operation of our electric appliances – and some is happening invisibly in the factories and offices that supply us with goods and services. In Europe, almost half of all energy use is a direct consequence of how individuals run their lives. The gas used to heat homes, the petrol for cars, jet fuel for airplanes, and the electricity for home appliances account for about 45 per cent of total energy use. Commerce uses another 20 per cent and industry accounts for the final 35 per cent.

Energy source	*Average continuous energy use per person (watts)*
Electricity	1,900
Other gas	1,400
Other coal	150
Other oil	1,650
TOTAL	5,100

The table above breaks down the energy footprint of the average person in Britain, a fairly typical advanced economy. A very small fraction of our current energy need is provided by renewable electricity sources. And about 15 per cent of the UK's electricity comes from nuclear power stations. But all the rest necessitates the burning of fossil fuels.

Of course, electricity is not itself a fossil fuel, but most power is generated from carbon-based sources, of which the most important are gas and coal. In the UK, almost 85 per cent of the coal and about a third of the gas that the country uses is burnt in power stations to make electricity.

Electricity

The proposals in this book will add to electricity demand. I estimate that the total production of power will need to rise by about 50 per cent to meet the extra requirements of electric cars and vans, the increased use of electricity for heating and cooling and the replacement of oil and gas in some industrial processes. (Many industries will continue to need fossil fuels for the most energy-intensive processes, such as smelting ores to make metals.) As the chart opposite shows, this will take electricity use to about 2,750 continuous watts per person, up from about 1,900 today.

How will this electricity be generated? By 2025, I believe it will be possible to have a mixed portfolio of renewable sources of power that provide most electricity without carbon emissions. Carbon capture technologies will stop the carbon dioxide from the remaining gas and coal plants reaching the atmosphere.

While these figures can only be guesses, how might this portfolio look across Europe?

Wind power	*25 per cent*
Solar power (mostly CSP in Africa)	*25 per cent*
Marine (tidal and wave)	*15 per cent*
Fuel cells and biomass CHP	*10 per cent*
Carbon capture (or nuclear)	*25 per cent*

Are these numbers feasible? Let's look first at wind power. If we want 25 per cent of the (increased) demand for electricity in Europe to come from wind, this means about 700 watts of continuous power per person, or about 6,000 kilowatt-hours a year. This will require a big wind turbine for every 1,800 people. If the turbines were in good

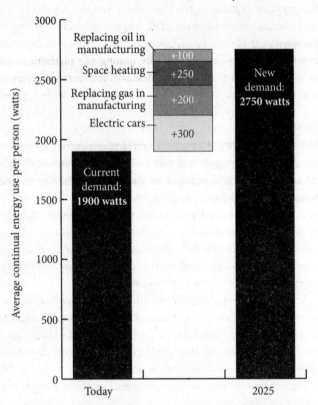

Possible increases in electricity demand

offshore locations and as large as the biggest of today's models, we'd still need one for every 3,000 people. That's 20,000 offshore wind turbines for a large country such as the UK, or as many as 35,000 if they were onshore.

Clearly this is a substantial challenge, but it is far from impossible. If all the turbines were put offshore, they would fill an area of about 100 kilometres by 160 kilometres. This is only a small percentage of the UK's continental shelf. It's worth mentioning that in some other parts of Europe almost 25 per cent of electricity demand is already met by wind.

Solar power, primarily produced in large-scale concentrated solar power plants in deserts, could also comfortably produce 25 per cent of Europe's electricity needs. As Chapter 2 demonstrates, the primary obstacle is the need for a sustained programme of construction of long-distance DC transmission lines. By contrast, the growth of electricity supply from tides and waves depends on continued entrepreneurial activity among the plethora of small firms constructing innovative devices. But for countries with long west-facing coasts a target of 15 per cent of electricity supply is very easily attainable. Meeting the fuel cell target also depends on technical progress, particularly in reducing the cost of cells designed to produce electricity for large commercial buildings. The use of biomass for combined heat and power needs few technological advances but is dependent on the ready availability of woody materials to use as fuel.

Generating this much power from renewable sources will require us to rethink the way we run our electricity grids. But, as discussed in Chapter 1, this is not an insuperable obstacle. We just need to develop more energy storage capacity, internationalise electricity grids, and find ways to reduce peak demand.

The remaining electricity demand is about 700 continuous watts per person, or 25 per cent of the total need for power. If, as is likely, we continue to use fossil fuels to generate this electricity, we will need to capture the carbon dioxide that results. CCS technology, as discussed in Chapter 8, is a long way from commercial availability. Indeed, it may be 2020 before we understand how to capture CO_2 with reasonable efficiency. But then it should be possible to add carbon capture equipment to most existing coal-fired power stations. The world may also decide to invest in large numbers of new nuclear power stations. In combination, nuclear and CCS-equipped conventional power stations will be able to produce the quarter of our electricity needs that does not come from renewable sources.

Other gas, oil and coal use

Completely decarbonising the production of electricity by 2025 is a tough challenge, but we can clearly see the routes that we need to take. Less work has been done on reducing gas demand. Gas that is not burnt to generate electricity is primarily used for heating buildings and to provide heat for industrial processes such as food

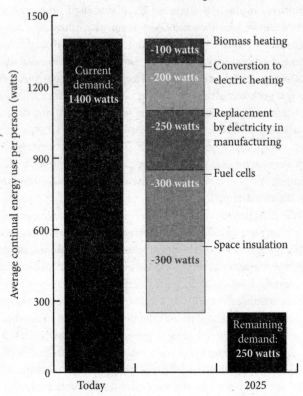

Measures to reduce gas demand

Average continual energy use per person (watts)

Current demand: 1400 watts

-100 watts — Biomass heating

-200 watts — Converstion to electric heating

-250 watts — Replacement by electricity in manufacturing

-300 watts — Fuel cells

-300 watts — Space insulation

Remaining demand: 250 watts

Today   2025

manufacture. In a country such as the UK, over half the gas used outside power stations is employed to provide heating for domestic homes.

Improvements in building insulation are the most effective way of reducing gas demand. The chapter on super-insulated homes shows that heat needs can easily be cut in half by relatively cheap improvements in insulation. Progress in cutting gas use in the home is fastest in places like Germany, where very substantial insulation improvements are benefiting several hundred thousand households a year. But even this figure means that only about half of a per cent of all existing households in the Federal Republic are undergoing major eco-renovation each year. The world needs to step up the rate at which insulation improvements are being made, but there are no technical obstacles to this, either in cold and in hot countries.

Further emissions reductions will come from increasing the amount of heating provided not by gas but by low-carbon electricity. We can also employ renewable fuels to provide heat, either through fuel cells powered by next-generation ethanol or by wood in district heating plants. Fuel cells for homes and for larger buildings will almost certainly be financially viable within a decade, particularly if today's gas prices persist, and the district-heating model has already been proven in Denmark and elsewhere.

Most oil is refined into diesel and petrol and used as a fuel for transport. In Chapter 6, we showed how liquid fuels can be replaced by electric batteries in cars and vans. Other vehicles can be powered by cellulosic ethanol, as described in Chapter 7. We will not find it as easy to replace kerosene aviation fuel or diesel for heavy vehicles. Although diesel substitutes can be made from any oil-bearing seed, biodiesel is problematic because its production generally involves the switch of either virgin forest or food-producing land to the growing of energy crops. We can hope that a large percentage of fossil fuel diesel can be eventually replaced by fuel made from dried algae, as described in Chapter 8, but the technology for this is not yet mature.

Oil currently provides about 1,650 watts of continuous energy in Britain and the figure is similar in many other industrial economies. About half of this energy is used in the fuel that drives cars and vans. The chart on page 261 shows how we can hope to substantially reduce the amount of fossil fuel but still use approximately the same amount of energy.

In this possible 2025 scenario, oil use falls by 600 watts as the electric vehicle takes over. (Electricity use only rises by about 300 watts because electric motors are more efficient at converting energy into motion than the internal combustion engine. The same amount of driving requires less energy.)

Even after the possible reductions detailed in this book, we are left with 650 watts of energy demand fulfilled by oil. This figure is composed largely of aviation fuel (about 300 watts), shipping (100 watts) and heavy road transport (200 watts). There is also a small

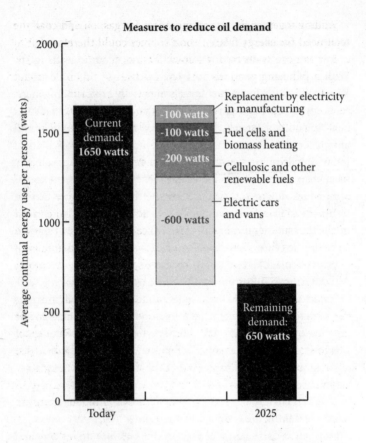

Measures to reduce oil demand

Average continual energy use per person (watts)

2000

Current demand: 1650 watts

1500

-100 watts — Replacement by electricity in manufacturing

-100 watts — Fuel cells and biomass heating

-200 watts — Cellulosic and other renewable fuels

1000

-600 watts — Electric cars and vans

500

Remaining demand: 650 watts

0

Today   2025

amount of industrial energy use that cannot easily be replaced with renewable sources.

Lastly, we need to consider coal. Excluding the use in electricity generation, coal provides about 150 continuous watts in a country such as the UK – only a small proportion of the total. Coal is generally used for heavy industrial purposes, such as in blast furnaces to make iron, and these uses will be largely unaffected by the technologies discussed in this book. The figure will also vary substantially country by country depending on the type of local industry. In some countries coal is also still used as a substantial provider of home heating. In these places, we can hope that coal will be replaced by renewable electricity.

Adding together the remaining needs for gas, oil and coal, the total need for energy from carbon sources could therefore fall to as low as 1,050 watts by 2025, down from over 5,000 watts today. With a following wind, the ten technologies in this book could therefore reduce fossil fuel energy needs in a typical advanced economy by up to 80 per cent. This figure conveniently matches most scientific assessments of the emissions reductions required in rich societies.

But we also have to consider the sources of greenhouse gases other than the burning of fossil fuels. Even if we reduce the use of carbon fuels down to about 1,050 watts, and therefore hugely reduce carbon dioxide emissions, we will still need to deal with the much smaller but still significant emissions of methane and nitrous oxide. To be safe, we might therefore choose to completely counterbalance the carbon dioxide from 1,050 watts of fossil fuel use by carbon sequestration in soils and plants.

The last two chapters of this book, on biochar and soil improvement, demonstrate that this is relatively easy and probably not expensive, though it will undoubtedly be difficult to organise on a large scale. Of equal importance to the climate change benefits, storing more carbon in the soil will probably improve the agricultural productivity of the world's land, increasing the amount of food that can be harvested and reducing the pressure to cut down the world's remaining forests.

How much extra carbon do we need to store in the soil to completely offset 1050 watts of fossil fuel energy so that the total amount of carbon dioxide in the air remains the same? This calculation is quite simple and very encouraging: 1,050 watts of continuous fossil fuel use implies a total of about 9,000 kilowatt-hours a year. If this energy were all generated by burning oil, this amount would mean using about 800 litres of oil per person, which would add about 2 tonnes of carbon dioxide to the atmosphere when burnt, containing somewhat less than 600 kilograms of carbon.

The chapters on soil improvement and biochar suggest that across large swathes of the world we can reasonably aim to increase the carbon stored in soils by at least 1 per cent of the weight of the soils themselves. If this were achieved, each person's 1,050 remaining watts of continuous fossil fuel could be offset by an area of just 10 metres by 10 metres per year. Of course, this has to go on year after year until we have completely decarbonised the economy, but that is perfectly possible. The UK, with less land per inhabitant than almost any other country in the world, could offset the emissions resulting from 1,050 watts of continuous fossil fuel use per capita for nearly thirty-five years before it ran out of space. In other parts of the world, the numbers are even more striking: Australia could sequester the carbon emissions resulting from its remaining fossil fuel use for several thousand years by achieving the same increase in soil carbon levels across the country. Increasing carbon levels in soils is also probably the cheapest way of reducing the carbon dioxide in the atmosphere.

A more intractable issue has been raised several times in this book. Sequestering carbon in the soil requires us to add plant matter, whether in the form of biochar, longer roots or greater amounts of humus. We also need to use wood and straw for making cellulosic ethanol for fuel cells and car engines and for providing the fuel for biomass heating plants. If the guesses in this chapter are approximately correct, we may need 800 continuous watts of our total per capita energy need to come from plant and trees. This is about the same amount of energy we might get from wind or solar power in 2025.

The photosynthesis process that creates all plant matter is much less good at translating energy from the sun into useful energy than, for example, a solar panel of similar size. To be clear, this means that we would need far less space if we employed photovoltaic panels to deliver our remaining energy needs than if we used trees to make ethanol or as fuel for a power plant. However, a hectare of solar panels will always be very much more expensive than a similar area of trees. The problem is that to get 800 watts of energy from wood and plant matter will require about an eighth of a hectare for each person in the industrial world. If a country such as France wanted to use its own land to grow the woody matter it needed, it would have to devote over 15 per cent of its total area to this purpose. Of course, it would probably make more sense to grow the wood or straw in less developed regions and then move the ethanol to where it is needed. This would have the benefit of increasing incomes in the developing world and providing a real incentive to reforest large areas. But it will still require a significant percentage of the world's usable land area to be given over to renewable forests for making the raw material for cellulosic ethanol.

It would be foolish to deny that developing a huge new industry that grows and processes billions of tonnes of forest matter around the world each year is a difficult challenge. Likewise, it would be foolish to think that decarbonising the world's electric grids is a simple task. But I believe – and I hope this book has shown – that it can be done.

Epilogue

The ten technologies, implemented and supported as a portfolio, possess the capacity to tackle climate change. Few of them can yet match conventional energy sources in terms of cost, even at today's elevated prices of oil, coal and gas. But we can be confident that energy produced by these new techniques will decline in cost and will eventually be competitive with fossil fuels. Similarly, the technologies that focus on energy saving or carbon sequestration will rapidly fall in price.

What is the basis for such optimism? Almost all manufacturing industries benefit from a learning curve. For example, at the point at which total wind turbine production exceeded 2 gigawatts the cost per unit of power was about 20 per cent lower than when the industry passed 1 gigawatt. This pattern of cost reduction is characteristic of almost all complex manufacturing and service industries. It is observed in aircraft manufacture, in semiconductors and even in repetitive machining operations or clerical tasks. (One rare exception, as we'll see below, is nuclear power-plant construction.) As industries become mature, the rate of decline in costs may fall, but the cost of almost everything we make tends to fall by at least 10 per cent every time cumulative manufacturing volumes double. This will be the case, for instance, in constructing biogas digesters on rural farms, building solar panels or improving home insulation.

A cost decline of 10 per cent may not sound much. Nevertheless it means that a technology growing by 40 per cent a year – a rate achieved, for example, by some types of solar power, by small fuel cells, and probably by electric cars – will nearly halve in cost over the next ten years. This almost certainly puts onshore wind power, cellulosic ethanol, probably solar power, and possibly some marine energy technologies in a position to be cheaper than natural gas or oil as a source of energy. Carbon sequestration will also go down a steep curve of cost improvement. If governments increase the price of fossil fuels by imposing carbon taxes or tight caps on their use, this price advantage will be even more apparent. And robust support for technologies such as the application of passivhaus techniques for existing buildings and the construction of cheap biochar kilns will also bring their price down. All the technologies in this book, including those that involve soil improvement, are at the beginnings of their life cycle and will be made significantly cheaper in the coming decades.

Politicians and investors need to have this point repeatedly made to them. All the ten technologies, with the possibly exception of wind, need support today, whether in the form of carbon taxes, explicit or covert subsidy, state sponsored research and development, or my own favourite, huge prizes for success. How should we reward, for example, the first company to successfully export a gigawatt-hour from the churning, underexploited tides of the Pentland Firth? How about with a £100m bounty and for the CEO a seat in the UK's upper legislative chamber, the House of Lords?

Omissions from the list of ten: what else could have been in?

A different writer may have picked a different list of ten technologies. I want to touch briefly on some of the other candidates: nuclear power, energy efficiency and geoengineering.

Nuclear power

With appropriate backing, nuclear power could comfortably provide most of the world's electricity needs in twenty-five years' time. (Its share is about 15 per cent today and falling.) So why isn't nuclear power one of the ten? Is this just another instance of a naive environmentalist irrationally opposing a well-understood science-based technology in favour of untried alternatives?

I acknowledge that many of the arguments against nuclear power are weak. One such argument is that there isn't enough uranium to go around. Despite recent enormous jumps in the price of refined uranium, the cost of fuel is unlikely to ever rise much above 1 US cent per kilowatt-hour of electricity output, a fraction of natural gas costs. Similarly, although current uranium fuel production levels are little changed from ten years ago, there is no likelihood of an early worldwide shortage of supply, even if a large number of new power stations are built. The extraction costs of the mineral may increase as the producers exhaust the mines with the richest sources of uranium, but the total amount of uranium ore available worldwide is likely to be sufficient for many decades of consumption. Supplies would tighten if the world added many hundreds of new reactors to the 400 or so operating today but uranium is never likely to run out. Although it is much more widely dispersed and only available in low concentrations, uranium is more common than tin in the earth's crust and it is even available in very dilute amounts in seawater.

Controversies over safety and the unresolved problem of how to store large quantities of nuclear waste remain. Recent leaks from reactors and processing plants in Britain and France show that although the safety record of nuclear power has been good there is no reason for complacency. The decommissioning costs of the existing generation of nuclear power stations will be high. This is particularly an issue in Britain, which is the only country in the world still using the earliest and rather crude designs for commercial reactors. Official estimates of the price for disposing safely of UK nuclear waste are rising, often by many billions of extra pounds every year. In addition, nuclear expansion almost certainly adds to the problem of the proliferation of weapons-grade radioactive material.

These particular problems may or may not be overcome but one further issue always remains. For reasons we do not completely understand, the world is very poor at constructing nuclear power plants on time and on budget. And it is getting worse. The chart on page 268 looks at the cost of building a nuclear power station in the US over the last few decades, expressed as the price the operator would have to charge to earn a reasonable return on the capital employed to build the plant. The upward drift in construction costs is obvious and some of the later nuclear power stations needed costs well in excess of 10 cents per kilowatt-hour to earn a reasonable return. This is far more than competing technologies. Many experts

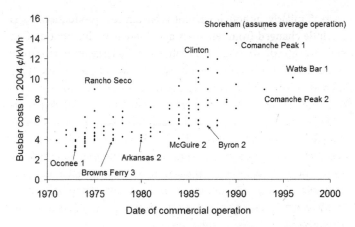

Construction costs for US nuclear reactors

expect companies such as First Solar to achieve a lower figure for solar panel installations by 2012.

Of critical importance, nuclear power stations have been getting more expensive as the technology matures, not less. This is in direct contrast to all the ten technologies chosen for inclusion in this book, which are all tending to get less costly the more that we make.

Recent years have seen relatively few new nuclear power stations. The nine built in Asia in the last fifteen years have cost an average of almost $3,000 per kilowatt-hour of capacity, if the cost is inflated to today's price level. This is far higher than the proponents of nuclear energy claim in their promotional materials. For example, the World Nuclear Association says the figure could be $1,500 a kilowatt-hour but provides no evidence to support this claim.

The price of building a nuclear station is the most important determinant of how much its electricity will cost to produce, so figures for construction costs are crucial to any understanding of whether nuclear power is genuinely competitive with the technologies discussed in this book.

The most recent example of construction cost overrun is a new reactor called OLK3, sited next to two existing nuclear plants at Olkiluoto in the west of Finland. Started in 2004, this project has experienced lengthy delays and may now not be ready until 2012 or later. The cost has risen substantially, possibly to double the initial estimates. The French construction company Areva is having to pay for the overruns and the eventual cost is likely to be about €6bn, or almost $6,000/$9,000 per kilowatt-hour, three times the price of the Asian reactors. This isn't an isolated example. Some US utilities are now openly forecasting that American nuclear plants might cost as much as $8,000 or $10,000 per kilowatt-hour even though construction costs are generally lower in the US. Indeed, many electricity companies are shying away from nuclear because of a belief that investors will simply not fund the risk of excessively high costs. This is a clear lesson from the international capital markets that governments in favour of nuclear power should not ignore.

All types of large infrastructure projects around the world, including the building of major gas and coal power stations, have tended to become more expensive over the last few years, but the inflation in the nuclear industry has been truly remarkable. In the 1970s, nuclear plants could be built for $1,000 or $1,500 dollars per kilowatt-hour at today's prices. Today the figure might be six times this level and there is no evidence that costs will eventually fall back to their earlier levels.

Are the nuclear cost increases a result of the inflationary effect of the prolonged boom of the early twenty-first century and the consequent shortage of spare capacity to design, build and manage major projects? Or is it because of greater public concerns about safety or the impact of these plans on the local ecology? Why does the world nuclear industry not follow the pattern of the learning curve seen in almost every other manufacturing process? The most convincing explanation for the sustained cost increases is that the nuclear industry has not managed to settle on a single design and then worked gradually to remove costs from this design by ironing out problems and inefficiencies. It might be said that the nuclear construction industry has never stopped building individual prototypes, all very different from each other. Nuclear construction costs are also increased by continuous enhancements of regulatory requirements and the need to spend huge sums to acquire permissions to build the power stations and to meet public objections.

The nuclear industry claims that it has now settled on three basic designs for the power stations to be built in the next two decades.

How the third nuclear reactor at Olkiluoto will look

It ought, at least in theory, to be possible to reduce the costs of these three types of reactor to well below the extraordinarily high levels seen in Finland at the moment. We will find out fairly soon whether any optimism is justified. China and other Asian countries will build large numbers of nuclear plants in the next decade. The Chinese national energy plan foresees that about a quarter of the increase in electricity consumption in the next twelve years will be provided by nuclear. If the country uses the Areva design now being so expensively pioneered in Finland, it will be building over eighty separate new plants. This cohort of power stations will be the conclusive test of whether the latest generation of atomic technology can create electricity safely and at a competitive price.

Chinese experience may well show that pessimism over nuclear construction costs is wrong. The published figures suggest that China expects construction bills of little more than $2,000 a kilowatt-hour, or less than a quarter of the prospective costs in Finland. If this Chinese figure turns out to be right, those Western countries which have turned against nuclear power will need to readdress their decision. But at the moment the financial case for investing in nuclear is strikingly absent. Rather than gamble $8bn or $12bn on a couple of nuclear power plants, private investors and government would be far better advised to back the renewable electricity technologies and carbon capture processes covered in this book. Or they could support alternative nuclear power sources such as the Thorium Molten Salt reactors advocated by Kirk Sorensen and his colleagues or the small, 'village-scale' plants being developed by Hyperion.

If reactors do actually cost $9,000 a kilowatt-hour, as implied by the Finnish experience so far, nuclear electricity is simply not competitive with the other main types of generation. Even if renewables fail to drop rapidly in price, it seems likely that, by 2020, coal-fired power stations with carbon capture will produce cheaper electricity than new nuclear plants. If we expect that to be true, there is no economic reason whatsoever for building nuclear generating plants. Since large numbers of new nuclear stations would also reduce the attractiveness of investing in competing 'baseload' generation from renewable sources, the arguments for supporting nuclear are further weakened.

Energy efficiency measures

Only one energy efficiency technology gets onto the list of ten – super-insulated homes. The lack of attention paid to other power-saving techniques perhaps needs a little explanation. House insulation is the single most important energy efficiency improvement the world can make, and it reduces both heating and air-conditioning needs. An individual in a cold country typically generates more carbon dioxide from home heating than from car travel. Other improvements, such as better refrigerators, or further improvements in lighting from the use of LEDs, cannot hope to provide anything like the scale of benefit that better home insulation offers.

A typical poorly insulated north European house might use 30,000 kilowatt-hours of heat a year. Among domestic appliances, the refrigerator usually provides most scope for improvement because new machines are very much better insulated than their equivalents of even ten years ago. However, replacing an old fridge with the best new model is unlikely to save more than 200 kilowatt-hours a year, less than 1 per cent of the energy consumed heating the house. Simply put, improving the amount of electricity used by home appliances is an important ambition, but not one which can hope to substantially reduce total energy demand.

Apart from poor housing insulation, the greatest source of waste in the consumption of energy is probably the internal combustion engine, which converts little more than a quarter of the energy in petrol into the kinetic energy which gets us from place to place. This book proposes that replacing conventional car engines with the much more efficient use of batteries and electric motors is the most sensible way of minimising this form of energy waste. So, strictly speaking, the chapter on electric cars also promotes an energy-saving technology.

After home heating and cars, the third most important target for efficiency improvements is the wasteful use of electricity in offices. The average office worker in the UK is responsible for about three times as much electricity consumption in a 40-hour working week as he or she is at home. Electronic appliances, such as computers and servers, are inefficient in their use of energy and are often kept on 24 hours a day. In summer, poor building design means that powerful air-conditioning is needed for the entire week, partly to remove the heat created by this electronic equipment. Reducing office electricity use should be a high priority, but rather than needing any new technology it simply calls for proper housekeeping and intelligent purchasing of low-energy appliances.

Geoengineering

If all else fails, can we avert global warming by emergency techniques to remove carbon from the atmosphere or block some of the sun's radiation reaching the earth? Some environmentalists rail against such 'geoengineering' schemes, saying that they encourage the world to continue with rash and unsustainable consumption of fossil fuels. Nevertheless, rational governments and scientific institutions must carry out research into this topic. We need emergency fallbacks in case emissions reductions fail or we find that temperature increases begin to induce dangerous instability in our weather systems. If, for example, glacial and ice cap melting begins to speed up dramatically, which paleoclimatic evidence suggests is a real risk, then even rapid emissions reductions will have no measurable impact. The thermal momentum of the ice means that melting will continue even if temperatures are stabilised. A large percentage of the world's population will face disaster as sea levels rise and the summer flow of glacier-fed rivers declines sharply, causing severe water shortages for hundreds of millions of people. In these circumstances, the only appropriate action is likely to be an attempt to rapidly reduce global temperatures.

The simplest way to reduce temperature is to cut the total amount of the sun's radiation reaching the earth's surface. Blocking 1 or 2 per cent of the solar energy that would otherwise reach us would be enough to counterbalance the effects of the greenhouse gases added to the atmosphere since the Industrial Revolution began. Two apparently viable techniques are canvassed for achieving this reduction – increasing pollution in the upper regions of the earth's atmosphere or inducing greater cloudiness in the lower regions of the air. We know from the eruption of Mount Pinatubo in 1991 that one violent volcanic eruption, blasting 20 million tonnes of sulphur dioxide 30 kilometres or more upwards, will provide enough of a solar umbrella to reduce temperatures by half a degree Celsius or more. In the case of Pinatubo, the effect lasted three or four years, changing weather patterns around the world, probably enhancing the drought in the Africa Sahel and causing excess rainfall in the US. Events on the scale of Pinatubo are rare: the twentieth century only saw one or two eruptions of equivalent size.

We could mimic the effect of large volcanic events by shooting sulphur compounds into the stratosphere. This idea has distinguished adherents, such as Nobel Prize winner Paul Crutzen, but most climate scientists are horrified by its potential side-effects. It might work at restraining temperature rises, but it would increase the rate of ozone depletion, change weather systems and increase acid rain. It would also do nothing to reduce the amount of carbon dioxide in the atmosphere, meaning that the oceans would continue to acidify as they absorbed increasing amounts of the gas. Among other effects, this will assist in the destruction of coral reefs and the gradual deadening of the seas arising from the loss of plankton and fish.

The Salter cloud-generating ship

Another way of reflecting sunlight is to create more low-level clouds. Paradoxically, wispy high-level clouds tend to keep heat in but thick layers of cloudiness near the surface send light back into space. Probably the most plausible way of increasing low cloud cover would be to create a fine mist of salty ocean water and spray it upwards. If done on a large and increasing scale, this would tend to increase the amount of cloudiness over the seas and help to decrease temperatures. One variant of this scheme is proposed by Professor Stephen Salter, the inventor of one of the early devices for capturing wave energy mentioned in Chapter 2. His plan is to have hundreds of automatically controlled wind-powered boats shooting spray into the air. But, once again, even if the plan works it doesn't reduce the bad effects of the increasing levels of carbon dioxide in the atmosphere. It simply masks more of the world's surface from the sun. Apparently bizarre other schemes, such as shooting trillions of tiny mirrors out into space to reflect sunlight, have similar flaws.

Others types of schemes for geoengineering try to increase the capacity of the seas to store carbon dioxide. The amount of carbon stored at the bottom of the oceans is many times what is in the air, the soil or trees. One idea for taking more carbon dioxide to the sea floor is to seed parts of the southern oceans with tiny iron filings. The theory is that the growth of plankton is held back by a shortage of iron, an important nutrient. Plankton absorb carbon dioxide in a photosynthesis-like process that produces calcium carbonate for their skeleton-like internal structure. When plankton die, they fall to the bottom of the ocean, carrying the carbon dioxide with them in the form of the carbonate. So increasing the number of plankton could help sequester carbon.

Experiments have shown that extra iron does indeed increase the growth of plankton. And since the plankton in the oceans have a total weight greater than all the trees and plants on the earth's land, this is potentially extremely useful. However, excitement was tempered when it was discovered that a very small percentage of the extra plankton actually fell to the bottom of the sea, where the carbon dioxide would be safely sequestered. What actually seemed to have happened was that bigger sea creatures simply ate more plankton, and were themselves consumed by species further up the food chain. Most of the extra CO_2 absorbed by the plankton eventually seems to have returned to the air. A similar scheme that involves sucking cold, nutrient-rich, deep-level water up to the surface in order to improve the rate of photosynthesis by tiny sea creatures may suffer from a similar problem.

Schemes such as these are widely derided by climate scientists, who tend to believe that geoengineering projects simply compound the original problem, rather than cure it. They correctly point out that mankind does not begin to comprehend many of the complexities of the world's weather and climate systems. To them, the idea that we could cleanly counteract the consequences of increasing greenhouse-gas levels by simple techniques such as reflecting the sun's energy is hubris of the worst sort. The great climate scientist Wally Broecker says that our increasing greenhouse emissions are having an effect on climate analogous to poking an angry beast with a sharp stick. Geoengineering may compound the risks by poking the animal with a second stick. It is far better to reduce emissions or increase the rate of carbon capture in the soil, plants and trees or by safe underground injection. (As we saw in Chapter 8, Broecker himself supports a scheme to chemically capture carbon dioxide from ambient air and sequester it underground: an approach that shouldn't have any climatic side-effects.)

But ruling out geoengineering entirely is surely a mistake. It is sensible contingency planning for a world that is only gradually waking up to the possible dangers of even modest further increases in atmospheric carbon dioxide levels. Although all geoengineering schemes will have risks, the possibility of unexpectedly rapid changes to the climate argues strongly for a sustained research effort. Even if we may never need to employ these techniques, we need to understand the best ways to reflect greater amounts of sunlight and improve the ability of the oceans to take up carbon dioxide without increasing acidification.

A carbon tax

The first pages of this book drew a distinction between those who think that mankind is doomed as a result of its voracious appetite for fossil fuels and those who believe that our capacity for technological improvement will eventually allow us to reduce carbon dioxide emissions without impairing growth in prosperity. I have tried to steer a course between these two increasingly polarised camps. I believe that the evidence is strong that the ten technologies in this book can substantially reduce fossil fuel use within a period of a decade or so. But to move rapidly to a point in which these technologies can have real impact, they need to be supported. The crucial requirement is a high and increasing price on carbon, imposed either directly on the carbon content of goods and services or through well-designed schemes that cap emissions but allow trading of permits. Both schemes mean that fossil fuels need to become more and more expensive, sending strong signals to energy users and encouraging them to use less coal, gas and oil. In addition, a high carbon price – perhaps $50 per tonne of carbon dioxide or more – would make almost all the technologies in this book competitive very soon.

The suggestion of a carbon tax of this size horrifies many free-marketeers, who claim it will impose extraordinary costs and cripple the economy. However, the maximum possible impact is about 1.5 per cent of the GDP of the typical rich country. Even this would only be the case if all the tax fed through into increasing energy prices but energy demand remained unchanged. Until last year, it seemed that the demand for fossil fuels was completely insensitive to their price. But 2009 saw encouraging signs that increased prices eventually affect the quantity of energy consumed. American gasoline demand is down and British domestic electricity use has also fallen as people have responded to the market price. A high carbon tax would encourage energy conservation measures, meaning that the net impact on the economy might be very small indeed.

Sensible taxation policy needs to keep the cost of carbon-based energy high and increasing, whatever the market price of oil or coal. An intelligent government will continue adding to the price of carbon fuels, while working to mitigate the impact on the poor, by such techniques as improving home insulation and subsidising the use of low-carbon energy. All the ten technologies in this book will move forward far more rapidly if innovators and entrepreneurs can be confident that the financial competitiveness of carbon reduction technologies will be consistently supported by governments. This is true both for electricity generation from wind, solar and marine sources and for carbon capture and storage techniques. A $50 per tonne carbon dioxide tax will help the world wean itself off fossil fuels and put in place huge programmes for carbon capture.

What does $50 a tonne mean in terms of costs to the consumer? Surprisingly little. Even the world's most polluting power stations, burning brown coal in old furnaces, generate about a kilogram of carbon dioxide for every kilowatt-hour that they generate. A $50 carbon tax would therefore add five US cents to the price of a kilowatt-hour. This is far from insignificant: at current British retail prices it might add 25 per cent to the prices householders pay for electricity. But it is not an economic catastrophe and most other uses of fossil fuels would see a much less significant change.

Scientists, entrepreneurs, activists and investors around the world have made huge progress towards solving the global warming problem through advances in technology. Governments across the world simply need to aid these people through intelligent and sustained support. In turn, electorates need to support those politicians who understand the need for coherent and sustained climate change programmes that last several decades. This last point provides the primary reason I wrote this book. I wanted to demonstrate to the inhabitants of democratic societies that the world's climate problems are probably solvable at moderate cost. We need to vote for governments that are prepared to take the somewhat painful measures, *today*, to permanently reduce our need for fossil fuels. Politicians who argue that climate change is too expensive to solve must be rejected – urgently.

Picture credits

US Department of Energy, p.6
Energy4all.co.uk, p.20
US Department of
Energy, p.23
Mariah Power, p.30
Siemens, p.33
MMA Renewable
Ventures, p.53
Polly Higgins, p.61
Gollmer/Solar
Millennium AG, p.63
Sandia National
Laboratory, p.65
Tec-uk.org.uk, p.67
gap-solar GmbH, p.139
Cailean Macleod, p.75
BioPower Systems
Pty. Ltd, p.79
Lunar Energy, p.83
MCT, p.85
Pelamis Wave Power, p.93
OpenHydro, p.86
Ceramic Fuel Cells
Limited, p.104
FuelCell Energy, p.108
Ulf Nilsson, Mim
Bild AB, p.117
H.G.Esch/Passivhaustagung.
de, p.120

ptiwin, p.125
Lars Pettersson/Scandinavian
Homes Ltd, p.130
Potton, p.135
Jay Premack, p.144
Tesla Motors, Inc., p.154
TH!NK, p.159
James Duncan Davidson, p.171
Wikimedia Commons, p.173
Range Fuels, p.175
Dr Stephen P. Long,
University of Illinois, p.183
Øyvind Hagen/
StatoilHydro, p.188
Vattenfall AB, p.199
HR Biopetroleum
Facility, p.202
BEST Energies, p.213
Rob Flanagan, p.215
BEST Energies, p.218
Wikimedia Commons, p.221
Dan Dagget, p.240
Li Kangmin/Mae-Wan
Ho/i-sis.org.uk, p.251
Jonathan Koomey and
Nathan Hultman, p.268
TVO, p.270
John MacNeill, p.274

Index